Logan Uriah Reavis

A change of national empire

Logan Uriah Reavis

A change of national empire

ISBN/EAN: 9783337118174

Printed in Europe, USA, Canada, Australia, Japan

Cover: Foto ©ninafisch / pixelio.de

More available books at **www.hansebooks.com**

OF

NATIONAL EMPIRE,

OR

ARGUMENTS IN FAVOR OF THE REMOVAL

OF THE

NATIONAL CAPITAL FROM WASHINGTON CITY

TO THE

MISSISSIPPI VALLEY.

(Illustrated with Maps.)

BY L. U. REAVIS.

Fair St. Louis, the future Capital of the United States, and of the Civilization of the Western Continent.—JAMES PARTON.

There is the East, and there is India.—BENTON.

ST. LOUIS:
PUBLISHED AND FOR SALE BY J. F. TORREY, BOOK AND NEWS DEALER.
1869.

Entered according to Act of Congress, in the year 1868,

BY L. U. REAVIS,

In the Clerk's Office of the United States District Court for the Eastern District of Missouri.

MISSOURI DEMOCRAT PRINT.

TABLE OF CONTENTS.

The Old Government, Statement and Map of.................... 9
The New Republic, Statement and Map of.................... 16
The National Growth and Material Power of the Continent 20
A Demand for a Change of the Seat of Government, and its
    Location at St. Louis.................... 44
The Geographical Argument.................... 47
The Population Argument.................... 49
The Commercial Argument.................... 53
The Political Argument.................... 61
The Conclusive Argument.................... 63
Special and Local Considerations.................... 165
What Time.................... 169

NOTICE.

WHILE in Washington City, last June and July, I talked with many persons in favor of the removal of the seat of government from that place to the Mississippi Valley. Before I left I was often met by citizens and visitors and questioned upon the subject. I made no disguise of my sentiments, but gave as my firm conviction that the seat of government would be moved, and that, too, at an early day. Talking with the Hon. Horace Greeley, on one occasion, upon the subject, he said that there was not a heathen city in the world as corrupt as Washington City, and that he was in favor of the Capital going anywhere to get it away from there. He jokingly added that he would never forgive the rebels for not taking Washington.

One day I was met by an old gentleman of ministerial proclivities, with whom I had conversed several times upon the subject. He said that many persons were making light of my project to move the Capital away from Washington; "but," said he, "I told them to not deceive themselves, that Noah preached one hundred and twenty years and the people would not believe him, but the flood did come as he had told them it would." Then said the old gentleman to me, "You keep at work, for a gimlet-hole will after a while sink a ship." I answered him that I most certainly should contend for the removal of the seat of government to that locality which was destined to hold the balance of power in the Republic.

Senator Sumner also expressed his belief that the Capital would be moved West, and that its removal was only a question of time.

One morning, while passing up Pennsylvania avenue, I was halted by an old gentleman who resides in Washington, and told that he understood I was there trying to move the Capital; I told him that he had been wrongly informed; that I was not there trying to move it, but was in favor of its being moved, and that I believed it would be moved. He asked me when; I told him in the course of five years. "Well," said he, I have lived here for thirty years, have studied the subject all over, and have never been able to see a single argument in favor of moving it." I said: "Sir, can you give me an argument to prove that the earth turns over?" He answered that he did not believe the earth did turn over; that it was all humbug to say that it did. I replied to him, saying that I could prove by astronomical argument that the earth did turn over, and that I could also give good reasons for the removal of the seat of government from Washington City to the Great West, but that I would not then give any arguments on either proposition.

I herein propose to give the arguments as intelligibly as I can in favor of the removal of the seat of government; nor shall I, in the attempt to give good reasons in favor of the change, try to deceive any American citizen by false reasonings, nor selfishly advocate the preference of any one locality for that great national purpose to the prejudice of any other place, but contend for that which I believe to be just and honorable in all relations to the Republic and her people.

In the preparation of these pages I do not claim by any means that I have exhausted the subject, but I have done the best I could with the material facts at my command. For my statistics I have consulted Government authorities and drawn freely from their pages. Especially am I indebted to the last Report (1867) of the Hon. Jos. S. Wilson, Commissioner of the General Land Office. This Report is no doubt the best that has ever come from that office.

While I claim correctness for my general statement of facts, I leave the reader to determine upon the weight of the argument and the merit of the subject under consideration.

Born and reared in the Valley of the Mississippi, and in a country and Government in extent and kind unequaled in the history of mankind, and sharing a little of that human nature which is keenly and instinctively alive to every step toward individual and national greatness, I cannot be otherwise than in favor of every change which our national progress demands. I am therefore committed to this work of removing the seat of government from the cradle of our national infancy to the Mississippi Valley, the destined home of our national greatness.

While writing my pamphlet entitled "The New Republic," I became fully satisfied that the special work of transferring the seat of empire from its present place to the Mississippi Valley would soon engage the attention of the greater portion of the American people, for the advanced column of civilization across the continent would demand the change as a fulfillment of the "Prophetic Voices about America."

THE WEST.

"Let her not be despised! American Orientals may dream that wisdom has taken up her perpetual abode on the shores of the Atlantic, and that the genii of Art, of Science, of Literature, have planted their rosy grottoes on the sunny side of the Alleghanies; but a thousand fancies never made one fact. Like the swaddled Hercules, the West has already put out her infant arms and strangled two political dragons that were coiling about her cradle; and as soon as she walks forth in the consciousness of matured strength she will make a greater fluttering among the harpies that prey upon her interests than did the club of the hero among the Stymphalian vultures. Ill-founded contempt is a blow that always rebounds. The Assyrians contemned the Persians while the Persians like muskrats were undermining the walls of Babylon. Haughty, learned, philosophic Greece, the conquerer of Xerxes, became a Turkish slave, and the fair daughters of Themistocles and Leonidas were bought and sold in the shambles of Smyrna. Rome despised the barbarians, and the barbarians conquered Rome. Cæsar overrun Gaul with victorious legions, and now Gaul holds a standing army in the city of the Cæsars. England would force America to drink Bohea, and America poured out for England a cup of gunpowder tea, the taste of which she has not yet got out of her mouth. Thus it is, Arms and Arts in their onward progress have always pitched their tents nearer the setting sun; and the conquests of the one and the triumphs of the other have left their fruits to ripen and decay on the track. The very relics of the ancient empires are now to be dug out of the soil. Civilization, like the ostrich in its flight, throws sand upon everything behind her; and before many cycles shall have completed their rounds sentimental pilgrims from the humming cities of the Pacific coast will be seen where Boston, Philadelphia, and New York now stand, viewing in moonlight contemplation, with the melancholy owl, traces of the Athens, the Carthage, and the Babel of the Western hemisphere."—*Horace Greeley.*

THE GREAT FIELD OF THE WEST.

"As the center of population and power is to be in the Mississippi Valley in the future, so must we look thither for the New Man who is to be the redeemer of our race and character. The Western man already shows larger, broader, and healthier development, spiritually speaking, than his brother of the East. He has never been cramped as yet by any of the restraining forms of social ecclesiasticism; his mind, like his eye, ranges over large extents, and is not content to sit down with itself after having acquired a little power over its fellows.

"As the Great West is bound to supply laws and men for the vast future of this continental country, so will it furnish the religion whose all-embracing forms are to invite the entire people into the simple secrets of its worship."—*Banner of Light, Boston.*

THE MEN OF THE WEST.

"One who has not visited the West knows little or nothing of the spirit of Western men. There is an all-pervading zeal, energy, ambition, push, go-ahead, seen nowhere else. The blood of a Western man courses more rapidly in his veins than in the Eastern man or in the European, and he thinks, talks, and acts on a larger scale. The Western farmer wastes more in a year than the Eastern farmer saves. He may lack refinement, but he has a generous heart for his friends and a deal of pluck for his enemies. His religion is less sectarian, less bigoted, and more broad, catholic, and truly christian."—*American Phrenological Journal, New York.*

"Now, sir, when I see this country, when I see its vastness and its almost illimitable extent; when I see the keen eye of capital and business fastened with steady, interested gaze upon the trade of the West, and all our Eastern cities in hot rivalry are reaching out their iron arms to secure our trade; when I see the railroads that are centering here in St. Louis; when I see this city with 60,000 miles of railroad communication and 100,000 miles of telegraphic communication; when I see that she stands at the head waters of navigation, extending to the north 3,000 miles and to the south 2,000 miles, and when I see that she stands in the center of the continent as it were; when I see the population moving to the West in vast numbers; when I see emigration rolling toward the Pacific, and all through these temperate climes I hear the tramp of the iron horse on his way to the Pacific Ocean; when I see towns and villages springing up in every direction; when I see States forming into existence, until the city of St. Louis becomes the center as it were of a hundred States, the center of the population and the commerce of this country—when I see all this, sir, I feel convinced that the seat of empire is to come this side of the Alleghanies; and why may not St. Louis be the future capital of the United States of America?"—*Extract from a speech of Senator Yates, of Illinois.*

"In whatever lands beyond the Sea the American citizen may sojourn, he carries with him the glowing sentiment of his country's greatness and capacity for mighty deeds. He carries with him its vast dimensions, as one would carry in his pocket a two-foot rule. He sometimes puts all the great rivers of Europe together between two banks, and measures against their united volume the giant Mississippi. He sketches the line of his country's length across the European continent, from the Pacific to the Mediterranean, and from the straits of Dover to the Bosphorus, and bids the by-standers note the results of the comparison. Now and then he demonstrates to the patriotic Briton how the whole of England might be put in Lake Michigan, leaving ample room for navigation on either side. Is the Frenchman or German proud of his native land, he suggests that both France and Prussia might be set down in the single State of Texas, and still leave territory enough within its boundaries to make a kingdom as large as Belgium."—*Elihu Burritt.*

THE SANGUINE.

"If it were asked whose anticipations of what has been done to advance civilization for the past fifty years have come nearest the truth—those of the sanguine and hopeful, or those of the cautious and fearful—must it not be answered that no one of the former class had been sanguine and hopeful enough to anticipate the full measure of human progress since the opening of the present century? May it not be the most sanguine and hopeful only, who, in anticipation, can attain a due estimation of the measure of future change and improvement in the grand march of society and civilization westward over the continent?"—*J. W. Scott.*

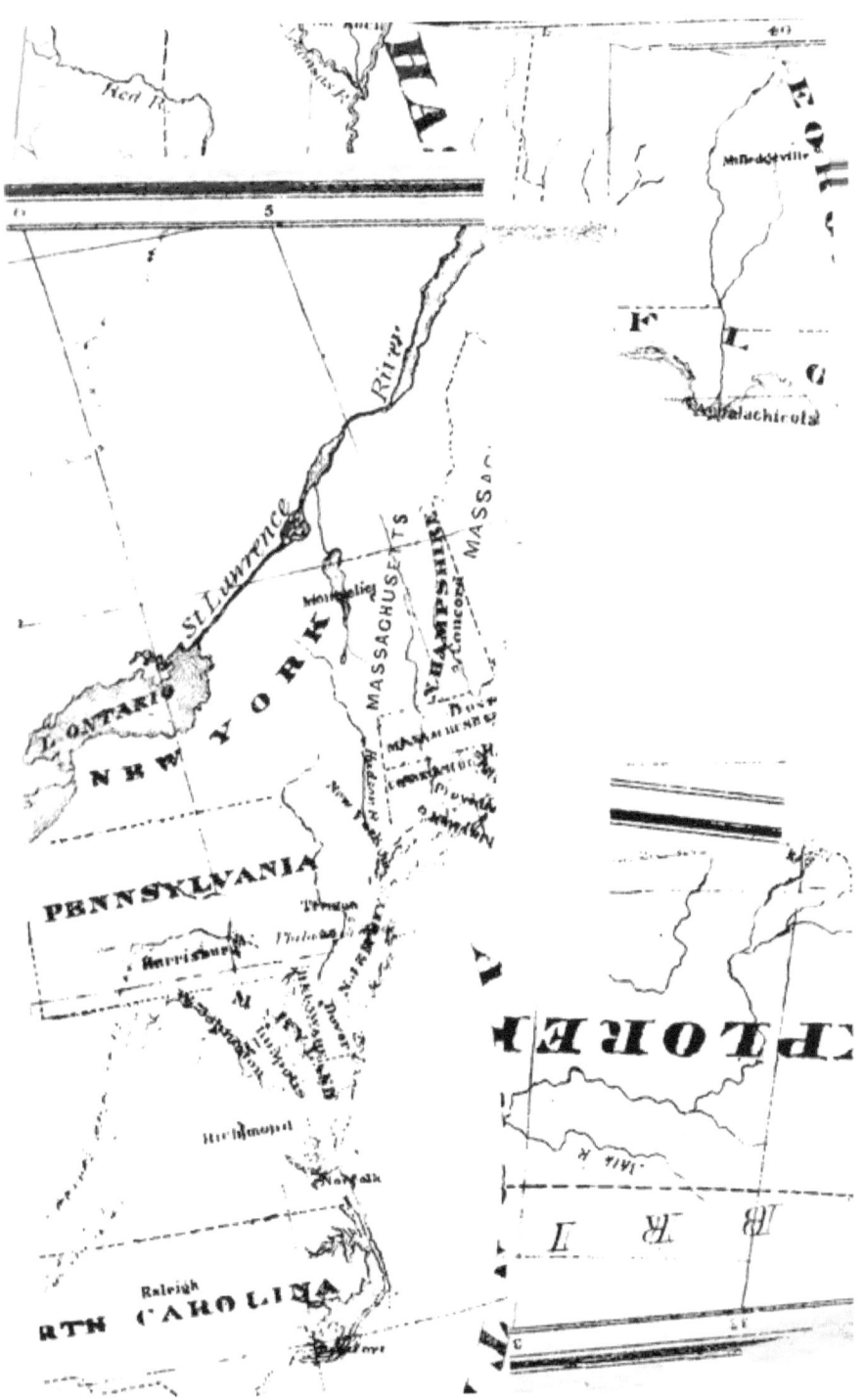

THE OLD GOVERNMENT.

But little more than the age of man, as assigned by the Psalmist, has passed away since the adoption of the Federal Constitution by our fathers, and the consequent creation of the infant government upon the Atlantic shore of the continent; and yet, in contrast with the living facts of the present, what we are to-day in power and greatness, the story of our national birth and growth seems but a romance—a mystic tale, told of the dim and shadowy past. History opens to our view back at our colonial period the most remarkable civil epoch in the career of mankind. We see by its light a strange people in a strange land struggling in a wilderness to found a new nation— they "builded wiser than they knew." When we contemplate that period, and know the newness of its history, it almost seems as if Washington had lived in the present generation; that Franklin, Jefferson, Paine, the Adamses, and Hamilton, had just ceased contending for human liberty, and had just founded "out of feebly-connected federal associations one people — an American nation." Venerable fathers and government-makers that they were, they have passed from mortal sight into everlasting history and heaven.

That the argument may be made stronger in favor of the removal of the National Capital from its present place to the Mississippi Valley, two maps of the country are submitted, with accompanying statements.

The first map represents the territorial extent of the United States Government at the time of the adoption of the Federal Constitution, and when the first Congress, sitting at New York, ocated the seat of government at its present place. In addition to the first map showing the territorial extent of the government at that period, it also shows the vast extent of wild country which has, since the incoming of the present century, been acquired by our government.

The first map represents the Old Government.

The second map represents the New Republic, or the territorial extent of the United States government as it now is, and in contrast with the Old Government we behold the growth of the American nation.

Let us turn back in our history eighty years, and briefly consider, in the interest of the subject of this pamphlet, what the Old Government was.

The following act locating the seat of government at its present place was passed by the first Congress, July 16, 1790, while in session at New York:

"AN ACT for establishing the temporary and permanent seat of the Government of the United States."—[1st Congress, Sess. II, Ch. 28, U. S. Statutes at Large.

SECTION 1. *Be it enacted by the Senate and House of Representatives of the United States of America in Congress assembled,* That a district of territory, not exceeding ten miles square, to be located as hereafter directed on the River Potomac, at some place between the mouths of the Eastern Branch and Connogochegue, be and the same is hereby accepted for the permanent seat of the Government of the United States: *provided,* nevertheless, that the operation of the laws of the State within such district shall not be affected by this acceptance until the time fixed for the removal of the government thereto, and until Congress shall otherwise by law provide.

SEC. 2. *And be it further enacted,* That the President of the United States be authorized to appoint, and, by supplying vacancies happening from refusals to act or other causes, to keep in appointment as long as may be necessary, three commissioners, who, or any two of them, shall, under direction of the President, survey, and, by proper metes and bounds, define and limit a district of territory under the limitations above mentioned; and the district so defined, limited, and located shall be deemed the district accepted by this act for the permanent seat of the Government of the United States.

SEC. 3. *And be it* [*further*] *enacted,* That the said commissioners, or any two of them, shall have power to purchase or accept such quantity of land on the eastern side of the said river, within the said district, as the President shall deem proper for the use of the United States; and, according to such plans as the President shall approve, the said commissioners, or any two of them, shall, prior to the first Monday in December, in the year one thousand eight hundred, provide suitable buildings for the accommodation of Congress and of the President, and for the public offices of the Government of the United States.

SEC. 4. *And be it [further] enacted,* That for defraying the expense of such purchases and buildings the President of the United States shall be authorized and requested to accept grants of money.

SEC. 5. *And be it [further] enacted,* That prior to the first Monday in December next all offices attached to the seat of Government of the United States shall be removed to, and until the said first Monday in December, in the year one thousand eight hundred, shall remain at, the city of Philadelphia, in the State of Pennsylvania, at which place the session of Congress next ensuing the present shall be held.

SEC. 6. *And be it [further] enacted,* That on the said first Monday in December, in the year one thousand eight hundred, the seat of the Government of the United States shall, by virtue of this act, be transferred to the district and place aforesaid, and all offices attached to the said seat of government shall accordingly be removed thereto by their respective holders, and shall, after the said day, cease to be exercised elsewhere; and that the necessary expense of such removal shall be defrayed out of the duties on imports and tonnage, of which a sufficient sum is hereby appropriated.

Approved July 16, 1790.

The following amendatory act was also passed by the first Congress, March 3, 1791, after the temporary removal of the seat of government to Philadelphia:

"AN ACT to amend an act for establishing the temporary and permanent seat of the Government of the United States."

Be it enacted by the Senate and House of Representatives of the United States of America in Congress assembled, That so much of the act entitled "An act for establishing the temporary and permanent seat of the Government of the United States" as requires that the whole of the district of territory, not exceeding ten miles square, to be located on the River Potomac for the permanent seat of the Government of the United States, shall be located above the mouth of the Eastern Branch, be and is hereby repealed, and that it shall be lawful for the President to make any part of the territory below the said limit and above the mouth of Hunting Creek a part of the said district, so as to include a convenient part of the Eastern Branch and of the lands lying on the lower side thereof, and also the town of Alexandria; and the territory so to be included shall form a part of the district, not exceeding ten miles square, for the permanent seat of the Government of the United States, in like manner and to all intents and purposes as if the same had been within the purview of the above recited act: *provided*, that nothing herein contained shall authorize the erection of the public buildings otherwise than on the Maryland side of the River Potomac, as required by the aforesaid act.

Approved March 3, 1791.

At the time of the passage of these acts there was not even a village where Washington City now stands, and, as will be seen by the act of July 16, 1790, the seat of government was not to be removed to its present place for ten years after the passage of the act, that time being given to prepare suitable accommodations, buildings, etc., for the transaction of business.

The following act, passed May 6, 1796, by the fourth Congress, while sitting at Philadelphia, provides for the public expense necessary to the permanent establishment of the seat of government at Washington City:

"AN ACT authorizing a loan for the use of the City of Washington, in the District of Columbia, and for other purposes therein mentioned."

SECTION 1. *Be it enacted by the Senate and House of Representatives of the United States of America in Congress assembled*, That the commissioners under the act entitled "An act for establishing the temporary and permanent seat of the Government of the United States," be and they are hereby authorized, under the direction of the President of the United States, to borrow, from time to time, such sum or sums of money as the said President shall direct, not exceeding three hundred thousand dollars in the whole, and not exceeding two hundred thousand dollars in any one year, at an interest not exceeding six per centum per annum, and reimbursable at any time after the year one thousand eight hundred and three, by installments not exceeding one-fifth of the whole sum borrowed in any one year; which said loan or loans shall be appropriated and applied by the said commissioners, in carrying into effect the above recited act, under the control of the President of the United States.

SEC. 2. *And be it further enacted*, That all the lots, except those now appropriated to public use in the said city, vested in the commissioners aforesaid, or in trustees, in any manner, for the use of the United States, now holden and remaining unsold, shall be and are hereby declared and made chargeable with the repayment of all and every sum and sums of money, and interest thereupon, which shall be borrowed in pursuance of this act; and to the end that the same may be fully and punctually repaid, the said lots, or so many of them as shall be necessary, shall be sold and conveyed at such times, and in such manner, and on such terms, as the President of the United States for the time being shall direct; and the moneys arising from the said sales shall be applied and appropriated, under his direction, to the discharge of the said loans, after first paying the original proprietors any balances due to them respectively, according to their several conveyance to the said commissioners or trustees. And if the product of the sales of all the said lots shall prove inadequate to the payment of the principal and interest of the sum borrowed under this act, then the deficiencies shall be paid by the United States, agreeably to the terms of the said loans; for it is expressly hereby declared and provided that the United States shall be liable only for the repayment of the balance of the moneys to be borrowed under this act, which shall remain unsatisfied by the sales of all the lots aforesaid, if any such balance shall thereafter happen.

SEC. 3. *And be it further enacted*, That every purchaser or purchasers, his or their heirs or assigns, from the said commissioners or trustees, under the direction of the said President, of any of the lots herein before mentioned, after paying the price and fulfilling the terms stipulated and agreed to be paid and fulfilled, shall have, hold, and enjoy the said lot or lots so bought, free, clear, and exonerated from the charge and incumbrance hereby laid upon the same.

SEC. 4. *And be it further enacted*, That the commissioners aforesaid shall semi-annually render to the Secretary of the Treasury a particular account of the receipts and expenditures of all moneys intrusted to them, and also the progress and state of the business, and of the funds under their administration; and that the said secretary lay the same before Congress at every session after the receipt thereof.

Approved May 6, 1796.

The District of Columbia, in which the seat of government is located, and which was defined in the act of July 16, 1790, was ceded by the States of Maryland and Virginia to the General Government. It consisted of a tract of country ten miles square until 1846, when by act of Congress (July 9 of that year) that portion ceded by Virginia was restored to her. The restoration was completed by a proclamation of President Polk, bearing date of September 7, 1846.

This left the Government in control of the portion ceded by Maryland, consisting of sixty square miles. The City of Washington was founded in 1793, and in 1800 the seat of government was moved from its temporary location at Philadelphia to its present place. At that time the entire territorial area of the United States was only 610,512 square miles, which was less than one-fourth its present size, exclusive of the Russian possessions; and by reference to the map of the country at that time it will be seen that the American nation, which our venerable fathers founded after years of toil and bloodshed, only consisted of a little strip of uninviting country stretching along the Atlantic shore from New Hampshire to Georgia, and the wild and unknown Northwestern Territory, reaching beyond the Alleghanies to the Mississippi river and the lakes. In other words, the United States at that time consisted of the following States and one Territory: New Hampshire, Massachusetts, Rhode Island, Connecticut, New York, New Jersey, Pennsylvania, Delaware, Maryland, Virginia, North Carolina, South Carolina, Georgia, and the Northwestern Territory. The population of the country at that time was 3,929,827, which was but little more than the present population of the State of New York. Our sea-coast was large, but our commerce of little value. We had not more than 1,000 miles of river navigation on the Atlantic slope, and the great lakes were far out in the West and of no use at the time. The Mississippi river was but little known, and even the Spaniards had navigated the Gulf of Mexico for two hundred years before its discovery. At that time there were no railroads, no steamboats, no telegraphs, but little education, and the continent still almost a wilderness, and our ancestors struggling against nature in her rudest form and the wild savages of the forests.

The debates upon the bill locating the seat of government at its present place show three considerations involved in the discussion:

First, that common selfishness which is everywhere seen in the acts of men. Many desired its location where it would build up local and personal interests.

Another argument was in favor of putting the Capital where it could be easily defended in time of war.

But the most important consideration was that which required its location in a central position, so as to accommodate the States as they were situated along the shore of the Atlantic. This, I repeat, as the debates upon the removal of the seat of government from New York to its present place show, was the most important consideration. The Constitution had just been adopted and the new Government took its place among the nations of the earth, and the representatives of the people at once sought to permanently locate the seat of government at such a place as would be most central to the States and the business interests of the people. Such was the wisdom of the representatives of the people at the foundation of the Old Government, and such ought to be the wisdom of the representatives of the people at the foundation of the New Republic.

THE NEW REPUBLIC.

Passing from a consideration of the Old Government, let us now turn to a consideration of the New Republic, or of our country as it is now, in all its broad extent. Little did our ancestors dream, when struggling for independence upon the narrow slope of the Atlantic, that they were founding a nation that would yet grow to be the greatest in mankind's history. Little did they know that they were organizing for a civil conquest of the continent — that from the parent home colonial columns would go out across the continent in every direction, seeking new homes and greater fortunes. No warrior ever prosecuted a conquest against any nation that conformed to more exact military rule than that of the civil conquest of this continent. While the central column was moving to the heart of the continent and onward to the great mountains, the right and left columns were flanking the great Lakes on the North and the Gulf on the South. Nothing retarded the movement or changed the direction of the pioneers and explorers but the arbitrary policy of the Government in establishing Indian reservations. While the central and right and left columns were marching in the front, a new movement was projected, and a force was sent around Cape Horn which entered by the Pacific Ocean, in the rear of the continent, on the golden shores of California; and thus the conquest goes on, and soon the will all meet and the continent be carved into one con. of great States.

By reference to the map it will be seen that the New Rep or the territorial extent of the Government as it now is, spans the continent in extent from ocean to ocean, and in breadth reaches from the lakes to the Gulf. Instead of the old thirteen States and one Territory, which constituted the Old Government,

the following new States and Territories have been added, which, in their broad extent and union with the old States, constitute the New Republic:

NEW STATES.

Kentucky,	Alabama,	Texas,
Vermont,	Maine,	Wisconsin,
Tennessee,	Missouri,	California,
Ohio,	Colorado.	Minnesota,
Louisiana,	Arkansas,	Oregon,
Indiana,	Michigan,	Kansas,
Mississippi,	Florida,	Nevada,
Illinois,	Iowa,	Nebraska.

TERRITORIES.

New Mexico,	Dakota,	Idaho,
Utah,	Arizona,	Wyoming,
Washington,	Montana,	N. W. America.

These added States and Territories in themselves combine all the elements of a great nation, far greater than that of our fathers. Whereas the area of the Old Government was 610,512 square miles, the New Republic has an area of 2,950,264 square miles, being more than four times greater in extent than the Old Government, exclusive of Alaska, which contains 577,390 square miles. With the expansion of the territorial extent of the Republic has also been added immense river, lake and ocean facilities for water transportation.

It is estimated that over two-fifths of our national territory is now drained by the Mississippi river and its tributaries, and more than one-half is embraced by what may be called its middle region, one-fourth of its total area belongs to the Pacific, and one-sixth to the Atlantic proper, one twenty-sixth to the Lakes, one-ninth to the Gulf, or one-third to the Atlantic, including the Lakes and the Gulf.

In reference to the facilities for water transportation, a calculation was made at the office of the Coast Survey, for 1853, which gives for the total main shore line of the United States, exclusive of sounds, islands, etc., twelve thousand miles, of which fifty-four per cent. belongs to the Atlantic coast, eighteen to the Pacific, and twenty-eight to the Gulf coast; and that if all these be followed, and the rivers entered to the head of tide water, the total line would be extended to 33,069 miles. Instead of 1,000 miles of available river navigation belonging to the Old Government, we now have in our broad extent about 20,000 miles, as follows:

	Miles.
Mississippi, from the Gulf of Mexico to Fort Snelling	2,131
Missouri, from mouth to Boseman	3,525
Ohio to Pittsburg	1,006
Illinois to LaSalle	300
Ouachita to Arkadelphia	601
Red River to Jefferson	720
Yazoo to Le Flore	257
Little Red to Searcey Landing	45
Arkansas to Fort Gibson	800
White to Forsyth	692
Black to Pocahontas	150
Currant to Doniphan	60
Tennessee to Florence	289
Cumberland to Nashville	193
Osage to Osceola	200
Kansas	200
Big Sioux	75
Yellow Stone	800
Minnesota	295
St. Croix	60
Chippewa	—
Monongahela to Geneva (slack-water, 4 locks)	91
Muskingum to Dresden, do 8 do	100
Green River to Bowling Green, do 5 do	186
Kentucky to Brooklyn, do 5 do	117
Kanawha to Gauley Bridge	100
Wabash to Lafayette	335
Salt to Shepherdsville	30
Sondey to Louisa	25
Rio Grande	2,000
Colorado	1,000
Sacramento	500
Columbia	500
Snake Fork	310
Clark's Fork	225
Willamette	200
Rivers of Atlantic Slope	1,000

NOTE.—Steamboats have ascended the Des Moines to Des Moines City, Iowa River to Iowa City, Cedar River to Cedar Rapids, and the Maquoketa to Maqaketa City, but only during temporary floods. Boats have gone up many other small rivers in past years, but as the country becomes more cultivated the wash and drift are greater, and the smaller streams fill up and are thus rendered useless for navigable purposes.

In addition to the immense increase of available river navigation, we have also acquired vast mineral fields of wealth in almost every part of our domain. So, too, have we added immense forests of valuable timber of all kinds necessary to supply the wants of the industrious and growing people.

Taking the continent as a whole, from the Atlantic to the Pacific, and from our northern boundary to the Gulf, it is not equaled in natural advantages by any country on the globe, and none other is more calculated to facilitate the advancement of civilization. Its immense navigable advantages, its dense forests of every variety of valuable timber, its outstretching expanse of fertile lands, and its inexhaustible and incalculable minerals, combine to make it the greatest nation of the earth in commerce, agriculture, mechanics, and wealth. In support of this statement, let us appeal to facts, and then see, after a careful examination, if we can judge anything of the future by the past.

Besides the immense acquirement of natural wealth, to us are given the wonderful creations of genius. We have the railroad traversing our lands everywhere; we have the steamboat upon all our navigable rivers; we have the telegraph connecting our cities, and the steam-engine doing our bidding in almost every phase of industrial enterprise. Thus we are, with all our continental growth, a new nation, requiring new laws, new advantages, and more appropriate uses in governmental affairs. Our unlimited sea-coast uniting us with all the commerce of the world, and our vast domain putting us within reach of every climate on the globe, and all our natural advantages combined, point to our future imperial greatness; and at every step we take forward wisdom tells us that the conditions and regulations of the Old Government are not adapted to the wants of the New Republic, for they were only the regulations and conditions of childhood, and not suited to the growth and maturity of manhood. It will be found, on examination, that we are met everywhere with evidence demanding a change of the National Capital from the Old Government to the New Republic.

THE NATIONAL GROWTH

—AND—

MATERIAL POWER OF THE CONTINENT.

Man everywhere and in all ages has ever sought for power and dominion. He has traversed the oceans, seas, continents, and islands; ascended the rivers and scaled the mountains; defied the climates and the great depths; and everywhere untiringly moves on after dominion and profit. Before our independence was achieved, the thought of continental empire had already entered the minds of many far-seeing persons in this and other lands. "Prophetic Voices about America" were not wanting in numbers to foretell the triumphs of that spirit of adventure which, in the fifteenth century, carried Vasco di Gama around the Cape of Good Hope, and Columbus to America. Even the age seemed to be instinctive with a better life, and prophets of one land and heroes of another were unqualifiedly pointing to America as the place for the future empire of the world.

As early as 1755, John Adams, but twenty years old, and the future statesman of Massachusetts, wrote to a friend in the following words: "Soon after the reformation a few people came over into this new world for conscience' sake. Perhaps this apparently trivial incident *may transfer the great seat of empire into America. It looks likely to me;* for if we can remove the turbulent Gallics, our people, according to the most exact computations, will in another century become more numerous than in England itself. Should this be the case, since we have, I may say, all the naval stores of the nation in our hands, it will be easy to obtain a mastery of the seas, and the united force of all Europe will not be able to subdue us."

This was the expression of a young school-teacher twenty-one years before the Declaration of Independence was made by the colonies. John Adams lived to see a system of government founded which, with broad and comprehensive policies, was destined to bring forth upon the American continent a nation of grander proportions and greater triumphs in civilization than his most enlarged understanding could comprehend.

His son, John Quincy Adams, at a later day, remarked of his father's letter: "Had the political part of it been written by the minister of state of a European monarchy, at the close of a long life spent in the government of nations, it would have been pronounced worthy of the united wisdom of a Burleigh, a Sully, or an Oxenstiern. *In one bold outline he has exhibited by anticipation a long succession of prophetic history, the fulfillment of which is barely yet in progress, responding exactly hitherto to his foresight,* but the full accomplishment of which is reserved for after ages."

Next to John Adams stands Mr. Jefferson, with clear conceptions of the future of the American nation. Soon after the treaty with the Kaskaskia Indians, by which was acquired a broad belt of territory extending from the mouth of the Illinois river to and up the Ohio, Mr. Jefferson first began to look with serious considerations to the future greatness of the nation; and the treaty with the Louisiana purchase led him to say that he "would not give one inch of the waters of the Mississippi river to any nation." And with prophetic conception he was again led to say, "When we shall be full on this side the Mississippi river we may lay off a range of States on the western bank, from the head to the mouth, and so, range after range, advancing compactly as we multiply." Thus it is, each succeeding generation does its work in fulfillment of the great prophecies of those wise men.

Before our independence was acknowledged, French Catholic missionaries had descended the Mississippi river, and by the right of discovery claimed the country along its shores for France, and named it Louisiana, after King Louis. In 1762 France ceded it to Spain. In 1800 Bonaparte became First Consul, and induced Spain to cede it back to France. Soon after the cession France became fearful of England on account of national difficulties, and sold the country to the United States for $15,000,000. This territory was known as the Louisiana Purchase, and included all the country of Louisiana, Mississippi, Arkansas, Missouri, Iowa, and part of Minnesota, Nebraska, and Kansas, besides a pretended claim to the whole territory extending to the Pacific Ocean. It must be kept in mind that all these vast possessions did not belong to the United States at the time of the location of the seat of government at its present place. In addition to the Louisiana purchase, Texas was annexed in 1845, New Mexico, California, and all the territory between the Mississippi river and the Pacific ocean has been added within the present century; and in rapid succession has State after State come into the Union, and the telegraph, the railroad, the steamboat, the printing-press, and the schoolhouse, have followed on in this great march of empire, and taken the place of the Indian trail, the wigwam, the hunting-ground, and the home of the buffalo.

Turn which way we will, upon this "vast, wide continent," and we see the chain of empire being made complete under one all-embracing Constitution. Climates of every character, minerals of every quality and value, rivers stretching in great lengths and uniting every zone, all combine to give greatness and destiny to this nation, made of the wisdom and excellences of all nations, and this people, made of the commingled and regenerated blood of all people. Sublime thought! Grandest and broadest of our age; that which energizes the individual and regales the future with royal promise.

Thus step by step has the Republic advanced in greatness, as predicted by the fathers, until with clear vision John Bright, the great English statesman and commoner, sees beyond the present the fulfilment of the prophecies of the fathers, and with conscious certainty speaks as follows of the future nationality of the New Republic:

"I see one vast confederation stretching from the frozen North in one unbroken line to the glowing South, and from the wild billows of the Atlantic westward to the calmer waters of the Pacific, and I see one people, and one law, and one language, and one faith, and over all that vast continent, the home of freedom and refuge for the oppressed of every race and of every clime."

COMMERCE OF THE OCEAN.

Turning from our national growth to a consideration of the material power of the continent, the first interest to be considered is the growth of our commerce upon the ocean. This element of our progress comes first, as the legitimate consequence of the infant nation having its existence along the Atlantic shore of the continent, which was akin to the commercial shores of Western Europe.

Besides the immense territorial expansion of our Government, its material growth and power is of infinite concern. The following tables show what our commerce was at the time of the adoption of the Constitution, and how it has grown until the present time.

But the growth of our ocean commerce has not been confined to our Eastern sea-board, nor to the development of the Atlantic slope, by any means. Already the Valley States furnish the greater part of the foreign commerce of the country, and the Pacific slope is also rapidly adding to its value and its tonnage.

A comparative view of the registered and enrolled tonnage of the United States, showing the tonnage employed in the whale fishery; also the proportion of the enrolled and licensed tonnage employed in the coasting trade, cod fishery, mackerel fishery, and whale fishery, from 1815 to 1866, inclusive.

YEARS.	Registered tonnage.	Enrolled tonnage.	Total tonnage.	Registered tonnage in the whale fishery.	Tonnage employed in steam navigation.	Coasting trade.	Cod fishery.	Mackerel fishery.	Whale fishery.
				Tons and 95ths.					
1815	854,294 74	513,833 011	1,368,127 78			435,066 86	26,510 33		1,229 92
1816	800,759 91	571,458 88	1,372,218 53			479,979 14	27,379 00		1,168 00
1817	809,724 70	580,186 66	1,389,911 41			481,487 92	53,960 26		349 92
1818	605,088 61	619,065 59	1,225,181 20	4,871 41		503,110 57	68,551 72		614 68
1819	612,930 11	647,821 66	1,260,751 63	16,151 77		523,536 30	65,041 92		686 33
1820	619,047 65	661,118 66	1,280,164 44	21,700 40		529,080 67	60,842 55		1,053 65
1821	619,896 60	679,062 30	1,298,968 70	25,391 44		559,435 57	61,351 49		1,920 40
1822	628,150 41	696,548 71	1,324,620 17	26,070 82		573,082 02	68,405 35		2,133 50
1823	629,920 76	659,614 87	1,289,565 68	45,449 42		566,498 88	67,621 14		585 37
1824	669,972 60	729,190 37	1,389,163 02	39,938 12		589,323 01	68,419 00		180 08
1825	700,787 68	722,322 69	1,423,110 77	33,465 70	28,879 03	587,972 07	70,026 02		
1826	737,978 15	786,212 68	1,524,190 83	35,979 24	21,061 02	606,429 44	65,761 42		216 83
1827	747,170 44	873,437 31	1,620,607 78	41,757 32	23,068 78	622,307 65	74,048 81		238 94
1828	812,619 36	928,772 51	1,741,391 87	45,653 21	40,197 55	738,922 12	74,947 74		180 34
1829	650,142 88	610,654 88	1,260,797 81	54,621 08	34,412 26	508,638 10	101,736 78		
1830	576,475 31	615,311 50	1,191,776 43	57,284 71	54,036 81	516,978 74	57,26,973 38		792 87
1831	620,451 22	617,334 21	1,237,816 25	39,911 82	62,471 74	529,723 74	60,967 81	48,210 80	481 42
1832	686,989 77	742,460 39	1,429,450 21	82,316 75	34,435 65	649,627 40	64,027 72	70.47, 421 72	357 45

1833	750,026 72	856,123 22	1,606,149 94	101,158 17	101,849 51	744,198 60	62,720 70	48,715 43	778 39
1834	857,438 42	901,468 67	1,758,957 14	108,060 14	122,815 02	783,618 65	54,403 70	61,082 11	364 16
1835	885,821 60	939,118 49	1,824,940 14	97,640 00	122,815 02	792,301 20	72,374 18	64,443 11	....
1836	897,774 51	984,321 14	1,882,101 65	144,680 50	145,556 39	873,023 21	62,307 37	64,425 25	1,573 26
1837	810,447 29	1,086,238 40	1,896,685 69	127,242 81	154,764 93	956,980 60	80,551 89	64,801 90	1,894 86
1838	822,591 86	1,173,047 89	1,995,639 80	119,629 89	193,413 58	1,041,105 18	70,064 00	56,649 16	5,229 55
1839	834,244 54	1,262,234 27	2,096,478 81	131,845 25	204,938 04	1,153,551 80	72,258 68	35,983 89	429 69
1840	899,764 76	1,280,999 35	2,180,764 16	136,926 61	201,339 29	1,176,694 46	76,035 65	28,269 00	....
1841	845,803 42	1,184,940 90	2,130,744 37	157,405 17	175,088 36	1,107,067 88	66,551 84	11,321 13	....
1842	975,358 74	1,117,031 90	2,092,390 69	151,621 71	229,661 18	1,045,753 39	54,804 02	16,096 83	377 31
1843	1,009,315 01	1,149,297 92	2,158,612 93	152,374 86	236,867 58	1,076,155 59	61,224 25	11,775 70	143 33
1844	1,068,764 91	1,211,330 11	2,280,095 07	168,293 63	273,179 33	1,109,614 44	85,224 74	16,170 66	321 14
1845	1,095,172 44	1,321,829 57	2,417,002 06	190,695 65	326,018 58	1,190,898 27	69,825 66	21,413 16	206 92
1846	1,130,286 49	1,431,798 32	2,562,084 81	189,980 16	347,893 02	1,298,870 89	72,516 17	36,463 16	439 58
1847	1,241,312 92	1,597,732 80	2,839,045 77	193,858 72	404,841 59	1,452,623 35	70,177 52	31,451 13	....
1848	1,360,886 85	1,793,155 00	3,154,041 85	192,179 90	427,891 03	1,620,988 16	82,651 82	43,558 79	432 75
1849	1,438,941 53	1,895,073 71	3,334,015 29	180,186 29	462,394 25	1,730,410 84	42,970 19	73,853 78	....
1850	1,585,711 22	1,949,743 01	3,535,454 23	146,016 71	525,946 90	1,755,796 42	85,646 30	58,111 94	....
1851	1,736,307 23	2,046,132 20	3,772,439 43	181,644 52	583,607 05	1,854,317 90	78,475 89	50,539 02	....
1852	1,890,448 20	2,238,992 27	3,138,440 47	173,793 77	634,240 67	2,008,021 48	102,659 37	72,546 18	....
1853	2,103,674 20	2,303,336 23	4,407,011 43	193,201 44	514,097 87	2,131,256 30	109,227 40	59,850 48	....
1854	2,333,819 16	2,469,083 47	4,802,902 63	181,901 02	676,607 12	2,273,900 48	102,194 15	35,041 14	....
1855	2,535,136 15	2,676,864 90	5,212,001 10	186,773 02	770,285 12	2,491,108 00	102,927 67	21,624 89	70 23
1856	2,491,408 63	2,380,249 73	4,871,652 46	189,213 29	673,077 54	2,211,935 45	95,816 08	29,886 54	247 73
1857	2,463,967 56	2,476,875 43	1,940,843 04	195,771 89	705,784 04	2,300,399 30	104,572 85	28,327 67	70 23
1858	2,499,741 79	2,550,066 51	5,049,808 35	198,593 51	729,498 41	2,361,595 72	110,896 44	29,593 80	....
1859	2,507,401 81	2,637,635 50	5,145,037 39	185,728 39	768,436 83	2,480,928 89	120,577 07	27,069 91	....
1860	2,546,237 09	2,807,331 33	5,353,858 42	160,841 30	847,957 49	2,644,867 08	136,653 70	26,110 50	....
1861	2,642,627 81	2,897,184 97	5,539,812 79	145,731 06	877,203 51	2,657,292 50	127,310 47	54,295 75	....
1862	2,291,251 19	2,820,913 29	5,112,164 48	117,713 64	710,462 33	2,568,545 73	122,862 82	80,596 50	....
1863	2,026,114 16	3,128,941 31	5,155,055 47	99,228 49	575,510 17	2,960,632 69	117,288 90	51,018 62	....
1864	1,581,894 47	3,404,505 32	4,986,399 79	95,145 38	978,177 74	3,204,227 32	92,744 86	55,498 40	....
1865	1,602,583 13	3,494,198 67	5,096,781 80	81,233 21	1,067,131 91	3,353,657 11	59,229 19	41,208 81	....
1866	1,492,923 85	2,817,851 63	4,310,775 48	43,070 20	1,083,511 99	2,689,151 10	42,796 18	46,589 46	....

AMERICAN AND FOREIGN TONNAGE ENGAGED IN THE FOREIGN TRADE OF THE UNITED STATES, 1789 TO 1821.

Years.	American, tons.	Foreign, tons.	Total tons.	Per-centage of foreign.	Years.	American, tons.	Foreign, tons.	Total tons.	Per-centage of foreign.
1789	127,329	106,654	233,983	45.5	1806	1,044,005	91,084	1,135,089	8.0
1790	354,767	250,746	605,513	41.4	1807	1,116,241	86,780	1,203,021	7.2
1791	363,662	240,548	604,210	39.8	1808	538,749	47,674	586,423	8.1
1792	414,679	244,278	658,957	37.0	1809	605,479	99,205	704,684	12.6
1793	447,754	163,566	611,320	26.7	1810	908,713	80,316	989,029	8.1
1794	525,649	82,974	608,623	13.6	1811	948,247	33,202	981,449	3.3
1795	580,277	56,832	637,109	8.9	1812	668,317	47,098	715,415	6.5
1796	675,046	46,846	721,892	6.4	1813	237,501	113,827	351,328	32.3
1797	608,078	72,757	680,835	10.6	1814	59,786	48,301	108,087	44.6
1798	522,215	87,790	610,005	14.3	1815	700,500	217,413	917,913	23.6
1799	624,839	107,583	732,422	14.6	1816	877,462	258,724	1,136,186	22.7
1800	682,871	121,463	804,274	15.0	1817	780,136	212,166	992,302	21.2
1801	849,302	157,270	1,006,572	15.6	1818	755,101	161,414	916,515	17.6
1802	798,805	145,519	944,324	15.4	1819	783,579	85,898	869,477	9.8
1803	787,424	163,714	951,138	17.2	1820	801,253	78,859	880,112	8.9
1804	821,962	122,141	944,103	12.9	1821	769,084	82,915	851,999	9.7
1805	922,298	87,842	1,010,140	8.6					

COMMERCE OF THE LAKES.

From the commerce of the ocean we pass to a consideration of the commerce of the great lakes. The following tables show their statistics of measurement and the tonnage of their carrying fleet:

TABLE OF MEASUREMENT OF THE LAKES.

Lakes.	Greatest length.	Greatest breadth.	Mean depth.	Elevation.	Area.
	Miles.	*Miles.*	*Feet.*	*Feet.*	*Sq. miles.*
Superior	355	160	900	627	32,000
Michigan	320	100	900	578	22,000
Huron	260	160	900	574	20,400
Erie	240	80	84	565	9,600
Ontario	180	35	500	232	6,300
Total	1,555	......	......	......	90,300

TABLE SHOWING THE CARRYING FLEET ON THE LAKES.

	No.	Tonnage.	Value.
Steamers	143	53,522	$2,190,300
Propellers	254	70,253	3,573,800
Barks	74	33,203	982,900
Brigs	85	24,831	526,200
Schooners	1,068	227,831	5,955,550
Sloops	16	667	12,770
Barges	3	3,719	17,000
Totals	1,643	413,026	$13,257,020

The above statement shows only the carrying fleet of the United States on the lakes. The Canadas also have a large fleet.

These lakes are estimated to drain an entire area of 333,515 square miles, and discharge their waters into the ocean through the River St. Lawrence, which is navigable from Lake Erie downward to all vessels not exceeding 130 feet keel, 26 feet beam, and 10 feet draft. Previous to 1800 there was scarcely a craft above the size of an Indian canoe in what was then called a pathless wilderness. The first American schooner launched upon Lake Erie was built at Erie, Pennsylvania, in 1797, but was soon lost.

The shipping employed on these great lakes has had various alternations of fortune. The development of steam and sailing vessels began to be conspicuous in 1833, and rapidly rose in the succeeding five years to 50,000 tons. In 1843 another great impulse was given to that trade, and, with the exception of a slight reverse in 1857, it has steadily increased to the present time. The present commerce of these lakes has an annual value of $450,000,000, or more than twice the external commerce of the whole country, and, as will be seen by the preceding table, is carried on by a fleet of 1,643 vessels.

COMMERCE OF THE RIVERS.

From the lakes and their commerce let us turn to the rivers and their commerce. I have already stated that the rivers of the Atlantic slope, the Mississippi and her tributaries, together with the rivers of Texas and the Pacific slope, would make 20,000 miles of navigable water. Upon these inland waters floats the greatest commerce of the country.

The following table exhibits the carrying fleet of the Mississippi and her tributaries:

Table showing the Conveying Fleet of the Mississippi river and its tributaries.

Ports.	Number of steamers.	Registered tonnage.	Carrying capacity.	Value in dollars.
*Cairo................	......	....	....	....
Cincinnati............	150	30,497.16	42,983	$4,134,000
Dubuque..............	20	3,204.37	5,137	459,500
Evansville............	25	3,043.51	5,019	402,000
Galena................	20	2,297.77	3,305	435,000
Keokuk...............	15	1,173.86	2,192	178,500
Louisville............	66	14,100.64	25,425	1,994,500
Memphis..............	70	9,849.62	15,121	1,011,200
*New Albany........	......	....	....	....
Nashville.............	12	1,183.06	2,156	108,000
*Natchez.............	......	....	....	....
New Orleans.........	80	15,860.07	21,625	1,292,000
Paducah..............	10	1,100.80	2,892	265,000
Pittsburgh (81 tugs)..	150	23,598.00	42,471	3,920,800
*Quincy..............	......	....	....	....
St. Paul..............	39	3,088.52	4,973	607,500
St. Louis.............	210	86,532.34	110,769	8,830,000
*Vicksburg...........	......	....	....	....
Wheeling.............	44	9,538.11	8,075	918,600
Totals............	910	216,067.83	292,144	$24,556,600

* No registration at these ports, for want of local inspectors.

Touching the commerce of the Mississippi Valley, I hereby submit a paper by Professor S. Waterhouse, of this city, which was read before the River Improvement Convention, held in St. Louis, February 12th and 13th, 1867. Although the letter has some parts not specially adapted to my purpose in this connection, on account of discussing outside interests, yet it contains that which bears with force directly upon this discussion, and will be found interesting to the reader:

Mr. President and Gentlemen of the Convention:

The right of a government to institute internal improvements is one of the essential incidents of sovereignty. Under all forms of polity, this power is justly vested in the central authority. Even despotic governments, which reverse the republican idea and administer affairs of state in the interest of a titled minority, have exercised this power for the benefit of the nation. Austria has expended large sums for the improvement of the navigation of the Danube. But a democracy rests upon the fundamental principle that the interests of the people are supreme. Our republican Government, in which is vested the exclusive control of internal improvements, is then bound by the most solemn obligations to consult the general welfare of the nation. But if it neglects this trust, then momentous interests which have been confided to its sole guardianship and fostering care must suffer, and popular rights, which can appeal only to constitutional processes of enforcement, will be ignored.

In the present instance, our duty is not arduous. The unmistakable jurisdiction of Congress, the frequent precedents and liberal policy of the Government, leave us only the easy task of showing that the proposed improvement of the Mississippi Rapids is a work of national importance.

The Mississippi and its affluents, draining an area of more than 1,000,000 square miles, and affording a water-carriage of more than 15,000 miles, form a system of river navigation unequaled in the civilized world. The entire coast line of the United States is less than 13,000 miles long; but the river line of the Mississippi and its tributaries, including both banks, is more than 30,000 miles long. The trade which now floats on these waters is immense. Its magnitude startles the imagination. In 1860 the total foreign commerce of the United States was $760,000,000. In 1865 the trade of nine cities on the Mississippi and Ohio rivers amounted to $747,000,000. The annual commerce of the Mississippi Valley is now estimated at $2,000,000,000. The yearly traffic of the *upper* Mississippi, which would be *directly* affected by the obstructions in the river, is $150,000,000. The amount of commerce which is annually deflected from the Mississippi by the difficulties of navigation is computed at $100,000,000. The yearly damage which the rapids inflict upon navigation is appraised at $10,000,000. In 1865 the direct loss occasioned by the impediments at Keokuk amounted to more than $500,000. The eight miles of obstructed navigation sometimes delay a steamer five days. This detention is a source of great expense. A steamer with a carrying capacity of 18,000 bushels of sacked grain would require a force of sixty hands. The daily cost of so large a crew is heavy. A delay of three or four days entails a great expense. After the improvement of the rapids, a tow-boat with the same motive power and a crew of twenty hands would transport 225,000 bushels of grain. The *Ajax* once towed from Louisville to New Orleans 460,000 bushels of coal. For more than half of the boating season navigation is embarrassed by low water on the rapids. During the period of shallow water no boat can carry freight enough for a profitable trip without lighting over the rapids. But the employment of barges involves a serious expense. In the absence of elevators it has necessitated the use of sacks. Wheat sacks now cost from seventy to eighty cents a piece; or, if hired, two and a half cents per bushel for each shipment. The expense of the four transfers at the Rock Island and Keokuk rapids is twelve cents a bushel, and the loss from waste is seven cents more. During the season of 1866 the Northern Line Packet Company paid $21,100 for lighting over the rapids. The packages received by this company numbered, in

1865..1,243,000
1866.. 979,000

This decrease of 264,000 packages was entirely due to low water. The company estimate their receipts for 1866, in case there had been uninterrupted navigation, at 2,500,000 packages.

The present method of handling grain is very expensive. The waste of grain by carriage in sacks, the extra labor, the transfer to the shore, the damage, the cost of tarpaulins, and the injury to the sacks, amount to 16 cents per bushel. The dangers of navigation increase the rates of insurance. The perils of the rapids add one-half of one per cent. to the price of every bushel of grain which is shipped to market from the Upper Mississippi. This assessment upon the industry of farmers is oppressive and unnecessary. Under all the existing difficulties the freight of cereals from the Upper Mississippi to New York is far cheaper by way of New Orleans than it is by the lakes and the New York canal. The comparative rates of transportation from Dubuque to New York are:

Via the lakes	68 cents per bushel.
Via New Orleans	38 " " "
Difference in favor of Southern route	30 " " "

The present cost of shipping grain from Chicago to Cairo by *rail*, and thence to New York by water, is no greater than the freight to the same point by way of the lakes. The existing winter tariff on wheat in bulk from Chicago to New York is:

By the lakes	44 cents per bushel.
From Chicago to Cairo, by *rail*	20 " " "
From Cairo to New Orleans, by water	12 " " "
From New Orleans to New York, by water	12 " " "

So great is the cheapness of river carriage that the rates of the Southern route, increased by 300 miles of costly railroad transit, do not exceed those of the Northern line.

There is an actual saving of 30 cents a bushel by the New Orleans route; yet at present, so great are the delays, risks, and infacilities of river transportation, the Northern lines of transit are still preferred.*

It is thought that, after the improvement of the rapids, the introduction of barges for the transportation and the erection of elevators for the transfer of grain in bulk, the freight of cereals from the Upper Mississippi to New York will be reduced to 25 cents per bushel. After the completion of these public works, the successful competition of the Mississippi would compel the railroads to reduce their rates of carriage. Even if there was no change in the channels of transportation, this reduction of freights would itself justify the removal of obstructions in the Mississippi. But there will be a change in the routes of freightage. Uninterrupted water carriage always affords the cheapest transportation. This fact is forcibly illustrated by the present movement of cereals. More than 75 per cent. of the grain received at Chicago is carried there by rail, but from that point only 10 per cent. is sent eastward by rail; 90 per cent. is shipped by the lakes.

It is sometimes alleged that the heat in the Gulf of Mexico is too great for the safe transport of grain by the Southern route. Corn is much more liable to be damaged by atmospheric influences than wheat, and the flour made from spring wheat is far more susceptible of injury from humidity than the grain from which it is manufactured. Yet the present trade of New Orleans in corn and spring-wheat flour is immense. Besides, the movement of Western cereals is made in the cooler months. Almost

---

* Since the original publication of this article, a reduction of the freights on Northern lines has diminished the relative cost of Eastward transportation, but there is still a difference of not less than 8 or 10 cents a bushel in favor of shipments to the Atlantic seaboard by the Southern route.

all our shipments of grain are made from September to June; so that, even if the midsummer heat of the Gulf was an objection to the Southern route, the difficulty would be obviated by the season of transportation.

The fact, too, that large quantities of Western flour are now exported without injury to the trans-equatorial countries of South America must not be ignored. Wheat is carried unharmed from San Francisco around Cape Horn to New York. The vast amounts of grain which are brought to Europe from the Danubian provinces, through the high temperature of the Mediterranean, reach their destination in a sound condition. The assertion, then, that cereals would be seriously injured by warmth and moisture in their passage through the Gulf is an allegation unwarranted by facts. A fear so foreign to commercial experience may be dismissed as a baseless apprehension.

But the Mississippi river, though entitled by a divine patent to the transportation of this valley, is now defrauded of its rights. An unlineal heir enjoys the inheritance. The value of the traffic deflected from the Mississippi into unnatural channels reaches an annual aggregate of tens of millions. In 1865, out of the 48,000,000 bushels of grain shipped to Chicago, 15,000,000 were brought from points on the Mississippi. According to Mr. Dodge, three-fifths of all the wheat received in 1865 at Milwaukee and Chicago came from the towns on the banks of the Mississippi.

The shipments were:

	Flour, bbls.	Wheat, bush.
East by rail	273,252	12,551,014
South by river	37,372	1,468,231

The following figures, furnished by Mr. Gilman, of Dubuque, express the actual cost of shipments from Chicago to New York:

Date.	Vessels.	Bushels.	Freights.	Sundries.
Oct. 1, 1865.	P. P. Cunningham	12,761	$4,608 01	$232 62
" 7, "	E. P. Dorr	11,679	5,527 25	552 85
" 21, "	Sailor Boy	18,700	7,946 87	445 02
" 31, "	Collingwood	16,313	6,634 56	495 95
Nov. 8, "	Dolphin	14,000	4,545 90	245 65
" 8, "	W. F. Allen	18,374	4,023 89	366 90
		87,827	$33,286 18	$2,438 00

There was also an additional charge of $2,195 67 at the Chicago elevators. Hence, the total expense of these shipments was $37,920 84, or more than 43 cents a bushel. This exhibit does not include commissions, storage, interest, insurance, government tax, or losses; but it does embrace wharfage, towing, measuring, sampling, and the cost of transfer at the Buffalo elevators.

These figures prove the supreme necessity of the projected improvements. The lakes are closed four months out of the twelve, but the Mississippi is open as high as Dubuque nine months in the year. Yet, notwithstanding this longer period of navigation and the continuous water carriage to Eastern markets, obstructions have almost wholly diverted the carrying trade of the Mississippi from its legitimate channel, and forced it into unnatural courses of transit. The unnecessary expense to which these impediments to navigation subject Western farmers is an oppressive tax upon agricultural industry. Agriculture is the basis of our public welfare. Upon it alone can rest an enduring superstructure of national prosperity. During the financial crisis of the last struggle, its unfailing resources alone upheld the credit of the public Treasury. Agriculture deserves the patronage of the Government. Its interests should be promoted by every aid of judicious legislation. But now, from the obstructions of navigation and the consequent want of competitive river transit, the railroad freight from the Mississippi to Lake Michigan costs more than one-fifteenth of the value of the grain. At the present price of wheat, this tariff, on the annual shipment of 50,000,000 bushels, would amount to $6,000,000. This yearly exaction is larger than the appropriation which Congress is asked to grant for the improvement of both rapids. The West now petitions Congress to grant relief from this hardship. An appeal sustained by such clear and imperative considerations of justice cannot be disregarded. A reduction of the cost of carriage is an object of national moment. It justly challenges the attention of statesmen; it affects the prosperity of the nation; it promotes alike the interests of the producer and the consumer; it enables the Western husbandman to make larger profits and buy more Eastern merchandise; it empowers the Atlantic manufacturer to live cheaper and sell more of his fabrics. The benefit is national.

At present almost the entire Eastern movement of cereals is carried on by way of the lakes. These Northern waters hold adverse possession of the carrying trade. The lake transportation companies have perfected all the machinery of freightage. They enjoy the advantages of long establishment, compact organization, and full equipment. But though the cost of shipment by the Mississippi is far less than by the lakes, adequate facilities for the transportation of our cereals do not exist on this river. There is no systematic combination, no means of conveyance commensurate with the wants of this valley.

But after the construction of the canals around the rapids, floating elevators and tow-boats will soon present ample facilities for cheap transfer and water carriage. Then the active competition of rival lines of barges and propellers will reduce still further the cost of Eastward shipments. This reduction in the rates of freights would be a national economy. It would lessen throughout the United States the expense of living. The quantity of Western cereals consumed in the Eastern States is immense. New England raises only one-fourteenth of the wheat which it consumes. Not even Pennsylvania and New York produce grain enough for their own consumption. All the Eastern and Southern States are largely dependent for their supply of flour upon the cereal products of the Mississippi Valley. In 1865 the receipts at the following points were:

	Flour, bbls.	Grain, bush.
Montreal	797,657	4,116,165
Portland	547,953	2,431,733
Boston	2,193,840	3,511,750
New York	3,687,775	37,339,903
Philadelphia	724,498	4,895,785
Baltimore	996,276	6,149,660
Tide-water by canal	1,014,000	45,830,100

After the deduction of our foreign exports of grain, the amount left for Eastern consumption is enormous. Diminish the cost of carriage, and you increase the profits and lighten the toil of every workingman in the land. Every mechanic, artisan, and operative in the Atlantic States would feel, in the amelioration of his condition, the beneficent effect of the contemplated improvements. The consummation of this work would enlarge the sales of every manufacturer in New England. The prime necessities of our national life are far more vitally affected by the unobstructed navigation of the Mississippi than by the security of our Atlantic harbors. Yet the Government has expended millions upon the improvement of the seaboard. Numerous and liberal appropriations have been made by Congress to insure the navigation of the lakes. Assuredly the Government cannot deny to our appeal the favor which it has granted to claims of no higher obligation. One year's interest on the value of the commerce which these obstructions divert from the Mississippi would pay for their removal. The annual tax which the rapids levy on Western products equals the estimated cost of the proposed canals. This valley is entitled to the cheapest transportation which unobstructed water carriage can afford All *additional* cost of transit is an unjust discrimination against agricultural industry. The difference in the price of grain between New York and the Mississippi Valley is a dead loss to the Western farmer. The heavy rates of freight levied on both eastward and westward exchanges oppress the producer with a double hardship. The cost of carriage is deducted from the value of Western grain, and added to the price of Eastern merchandise. This two-fold grievance, of which the West so justly complains, ought at once to be redressed. Congress should confer the earliest and the fullest relief which the nature of the case permits. An adherence to its settled policy, fidelity to its responsible trusts, and its high obligation to recognize popular rights and foster national interests, urge the Government to grant the solicited appropriation.

Thus far, our attention has been mainly occupied with the consideration of a single interest. But the completion of this public work would not only affect the cereals, but every other product of the West. While it would encourage agriculture with larger rewards, it would stimulate all industries by fostering the source of their common prosperity. It would invest the Mississippi with its rightful control of the heavy exports and imports of this valley. It would develop commercial activity, and greatly promote the interchange of productions between different latitudes. It would hasten the return of the South to its true allegiance, and bind it to the Union with the strong ties of sectional interest. It would augment our foreign commerce. It would favor the *direct* exchange of heavy commodities. In 1862, more than 80,000,000 bushels of grain, including flour, were exported from the United States. Though the effect of civil war upon our foreign commerce was disastrous, yet the value of breadstuffs exported from this country during the five years ending with 1865 was more than $360,000,000. If the United States possessed that control of European markets which the improvement of the Mississippi and the consequent cheapness of exportation would secure, our shipments of breadstuffs would expand into far grander proportions. The profits which the Atlantic cities would derive from this enlargement of our foreign commerce is an additional reason why the East should strenuously co-operate with the West to secure the consummation of this great work.

But the West has a higher title to the favor of the Government than the consideration of mere material interests. Faithful to its patriotic instincts, the West fought for the Union throughout the late contest with a stubbornness of valor that was at once a defiance of defeat and a guarantee of victory. Without disparagement to the noble gallantry of Eastern soldiers, it was chiefly due to the heroic efforts of our Western armies that the Mississippi now flows free to the Gulf. Their dauntless courage prevented the rupture of our national integrity, and rescued the mouth of the Mississippi from the control of a foreign power. Their fidelity has saved the Mississippi from the vexations of hostile imposts, and permitted its waters to flow untaxed to the ocean. To their service is to be ascribed the restoration of that unity and brotherhood for which the plastic hand of nature channeled this majestic stream. Assuredly the nation cannot forget its defenders. A government justly sensible of its obligations will show a practical gratitude for the preservation of its life.

The laws of trade ultimately enforce obedience. The imperial Mississippi, which traverses the central valley of this continent, and, independent of its tributaries, washes the borders of ten States, will yet assert its commercial sovereignty. The God of nature has invested this majestic stream with rights of conveyance which no railroad powers of attorney can transfer. The title of the Mississippi river to the commerce of this valley is attested with the Divine signature. The productions of the West will be borne to the tide-water through channels which the Architect of nature formed. Our Western rivers will soon transport a greater wealth of traffic than ever before floated on inland waters. The usefulness of the projected improvement will increase with the growth of the Mississippi Valley.

The following table shows the population and grain crop of the eight Northwestern States during the last three decades:

Years.	Population.	Bushels.
1840	3,340,500	165,698,800
1850	5,403,600	210,950,300
1860	8,855,900	536,801,900

The Agricultural Bureau, basing its calculations on past results, makes the following approximate estimate of the cereal product of the Northwest for the next four decades:

Years.	Bushels.
1870	762,200,000
1880	1,219,520,000
1890	1,951,232,000
1900	3,121,970,000

These numbers indicate a vastness of agricultural production and commercial exchange which the mind fails to grasp. Our conceptions of the future greatness of the West are rather embarrassed than aided by these figures. In the coming time, tens of millions will throng this valley under the benign sway of one government. All the prosperities of a free people and a Christian civilization will gladden this land. Our waste territories will become populous States. The resources of the soil and mine will be developed. Our wealth of agricultural and mineral productions will enrich the world. In that day the Mississippi will bear upon its bosom a commerce richer than the golden freights of classic story, and vaster than the maritime trade of any people on the globe. Our Government ought at once to prepare the Mississippi for its glorious destiny.

RAILWAYS.

Passing from our commerce upon the ocean, the lakes, and the rivers, let us turn to a consideration of our vast system of railway—the wonderful creation of American genius, industry, and wealth.

Table showing the Annual Progress of Railways for the last forty years.

Year.	Miles.	Year.	Miles.	Year.	Miles.
1828	3	1842	3,877	1856	19,251
1829	28	1843	4,174	1857	22,625
1830	41	1844	4,311	1858	25,090
1831	54	1845	4,522	1859	26,755
1832	131	1846	4,870	1860	28,771
1833	576	1847	5,336	1861	30,593
1834	762	1848	5,682	1862	31,769
1835	918	1849	6,350	1863	32,471
1836	1,102	1850	7,475	1864	33,800
1837	1,421	1851	8,589	1865	34,442
1838	1,843	1852	11,027	1866	35,361
1839	1,920	1853	13,497	1867	38,000
1840	2,197	1854	15,672	1868	41,358
1841	3,319	1855	17,398		

Table showing the Railways of the United States by States.

UNITED STATES.	Miles.	Cost.	Cost per Mile.	Area of Country.	Population 1860.	Miles of R. R. to Sq. Mile.	Miles of R. R. to Population.
				Sq. Miles.		Sq. M.	Pop'n.
Maine	502.37	$18,242,215	$36,315	31,766	628,279	62	1,234
New Hampshire	559.33	22,052,063	35,446	9,280	326,073	14	435
Vermont	591.59	24,892,231	41.864	10,212	315,098	17	529
Massachusetts	1,350.86	79,406,774	55,704	7,800	1,231,066	6	925
Rhode Island	119.24	4,858,799	40,787	1,306	174,620	11	1,467
Connecticut	647.54	24,370,018	38,225	4,674	460,147	7	721
New York	3,025.39	152,570,769	50,431	47,000	3,880,735	15	1,283
New Jersey	901.41	55,204,403	61,913	8,320	672,035	9	743
Pennsylvania	4,037.15	210,080,309	52,037	46,000	2,906,115	11	720
Delaware	157.49	5,604,864	37,279	2,120	112,216	13	714
Maryland and D. C.	522.60	30,573,275	58,591	11,184	702,129	21	1,457
West Virginia	364.75	24,978,843	68,408	23,541	349,028	56	958
Kentucky	625.90	22,392,122	35,776	37,680	1,155,684	60	1,846
Ohio	3,402.98	135,231,975	39,739	39,964	2,339,511	11	687
Michigan	966.12	41,675,724	43,133	56,243	749,113	58	775
Indiana	2,211.80	79,186,767	35,802	33,809	1,350,428	15	610
Illinois	3,250.05	139,081,414	42,791	55,405	1,711,951	17	527
Wisconsin	1,045.41	40,081,369	38,343	53,924	775,881	51	742
Minnesota	392.00	12,450,000	31,760	83,531	172,123	213	439
Iowa	1,154.10	45,480,000	39,407	55,045	674,913	47	498
Missouri	957.75	51,357,077	54,935	67,380	1,182,012	72	1,260
Kansas	240 50	9,750,000	40,540	78,418	107,206	327	445
Nebraska	275.00	12,500,000	45,454	76,928	28,841	279	105
California	321.50	21,200,000	75,272	188,982	379,994	508	1,180
Oregon	19.50	500,000	25,641	95,274	52,465	5,014	2,690
Virginia	1,416.70	40,974,457	35,275	61,352	1,246,381	43	879
North Carolina	977.30	20,020,310	20,485	50,704	992,667	52	1,016
South Carolina	988.93	25,297,977	25,491	29,385	703,812	28	711
Georgia	1,457.22	29,177,663	20,301	52,000	1,057,329	36	737
Florida	407.50	8,808,000	21,762	59,269	140,439	145	345
Alabama	891.16	21,010,982	25,154	50,722	964,296	57	182
Mississippi	867.12	25,416,391	29,315	47,156	791,336	54	913
Tennessee	1,316.73	34,185,210	25,937	45,600	1,109,801	34	842
Arkansas	191 00	4,400,000	43,562	52,198	435,427	273	2,279
Louisiana	355.75	13,637,654	40,577	46,431	708,290	138	2,111
Texas	479.50	17,280,000	36,044	237,504	602,432	495	1,257
Territories				1,243,416	524,387		
Total	36,806.26	$1,517,510,765	$41,120	3,001,002	31,747,514	81	860

By the preceding tables it will be seen we have in the United States 36,896 miles of railway, built at an expense of $1,517,510,-765. All this has been done since the adoption of the Federal Constitution, and since the location of the seat of government at its present place. But this is not all; their construction over the country, and especially Westward, is being pushed forward at the rate of fifteen miles per day, and next year the two great oceans bounding the Republic will be united by the completion of a great railway across the continent; and with our frontier line advancing fifteen miles every year, the civil conquest of the continent will soon be complete. But, in justice to the cause of the subject of this pamphlet, let it be said in truth that these mighty works are being done in the Mississippi Valley and westward of the Father of Waters.

Since the invention of the steam-engine, the railway system may be regarded as the greatest aid to civilization the arts have afforded, on account of the rapid intercommunion of men and ideas, and the exchange of products. Every additional investigation by the political economist and the socialist proves the influence of the railway upon the industry and intelligence of man to be the most potential of all his works. And it does really appear that the use of the railroads is destined to make all the agricultural interests of men subserve their highest uses, by enabling the producer to get the highest possible price for his produce, and the consumer at the least cost. The influence of railroads upon agriculture has been ably discussed by Mr. Joseph C. G. Kennedy, Superintendent of the United States Census for 1860, in the Agricultural Department of the Eighth Census. Speaking of their great value, Mr. Kennedy says:

"So great are their benefits, that if the entire cost of railroads between the Atlantic and Western States had been levied on the farmers of the central West, they could have paid it and been immensely the gainers. This proposition will become evident if we look at the modes in which railroads have been beneficial, especially in the grain-growing States. These modes are—first, in doing what could not have been effected without them; second, in securing to the producer very nearly the prices of the Atlantic markets, which are greatly in advance of what could have been got on his farm; and third, by thus enabling the producer to dispose of his products at the best prices at all times, and to increase rapidly both the settlement and the annual production of the interior States."

Mr. Kennedy gives the following table, showing the cash value of farms in five States, with their increase in ten years:

	1850.	1860.
Ohio	$358,758,602	$666,564,171
Illinois	96,133,290	432,531,072
Indiana	136,385,173	344,902,776
Michigan	51,872,446	163,279,087
Wisconsin	28,528,563	131,117,082
Aggregate	$671,678,075	$1,738,394,188
Increase in ten years		$1,066,716,113

Mr. Kennedy says it is not too much to say that one-half of this increase has been caused by railroads.

But the beneficial influence of the railroads cannot be confined to agriculture alone. Their influence is immeasurable upon the development of every commercial and industrial movement of our people, and consequently aids vastly the increase of population; and with the unequaled advantages for their construction and their use in the Mississippi Valley, they must be accounted a great auxiliary to the internal development of material power on the continent, and consequently of establishing the supremacy of the States of the Mississippi Valley over those of both oceans, thus giving to them that supremacy in civilization which is theirs by nature.

No new field of art or industry now engages so much capital, and is pushed forward with so much enterprise, as that of the railroad interest of the country. Where they are not, stagnation in business and conservatism in public spirit prevail; where they are, commerce and industry are vitalized.

To show the preponderance of material power and wealth in the Mississippi Valley, the following table is submitted. It exhibits the growth of our people, their genius, their wealth, and their wonderful industry:

STATISTICAL TABLE,

Prepared from United States Census Reports, showing the Controlling Power and Progress of the Mississippi Valley.

DESCRIPTION OF RESOURCES.	ATLANTIC SLOPE. 1850.	ATLANTIC SLOPE. 1860.	MISSISSIPPI VALLEY. 1850.	MISSISSIPPI VALLEY. 1860.	PACIFIC SLOPE. 1850.	PACIFIC SLOPE. 1860.
Population	18,255,254	15,993,302	9,813,117	14,908,427	107,271	491,153
Area, square miles	423,197	423,197	1,899,811	1,899,811	627,256	627,256
Land improv'd in farms, acres	43,965,491	73,882,853	48,985,479	87,034,199	181,614	3,537,668
Land unimprov'd in farms, acres	84,358,354	93,679,468	39,726,948	142,567,254	4,191,998	7,763,030
Cash value of farms	$1,991,569,478	$3,132,561,500	$1,232,941,038	$3,416,702,533	$6,033,010	$47,780,334
Value of farming implements and machinery	$78,826,805	$105,820,439	$72,339,639	$134,292,513	$371,194	$3,985,091
Live Stock, Horses	1,411,417	2,054,260	2,460,078	3,987,645	32,194	207,260
Asses and Mules	730,785	270,187	326,128	641,056	2,411	5,805
Milch Cows	3,445,181	3,194,557	2,421,315	4,450,022	18,568	281,151
Working Oxen	816,236	755,094	816,348	1,393,995	18,160	45,832
Other Cattle	4,619,672	4,615,368	5,372,321	9,099,605	280,276	1,075,314
Sheep	10,665,773	8,701,365	11,043,228	12,644,001	36,218	1,221,019
Swine	10,167,505	9,767,182	20,152,783	23,254,291	33,925	551,672
Value of Live Stock	$280,479,888	$382,675,639	$257,926,413	$617,616,940	$5,774,215	$44,325,528
Wheat, bushels	47,636,178	53,396,807	48,474,551	102,057,361	336,973	7,229,988
Rye, bushels	12,805,283	15,287,195	1,383,214	4,958,375	316	65,844
Indian Corn, bushels	180,029,565	201,633,963	350,912,515	636,456,535	25,033	682,486
Oats, bushels	82,365,111	93,216,624	62,646,947	72,350,255	72,143	2,130,306
Rice, pounds	205,439,416	179,427,085	9,874,081	7,730,807		2,140
Tobacco, pounds	94,102,203	215,472,069	105,646,057	218,732,827	1,395	3,565
Gin'd Cotton, bls 400 lbs each	889,615	1,278,646	1,546,178	4,108,270		136
Wool, pounds	27,881,509	25,302,154	24,381,022	30,765,734	44,428	2,997,035
Peas and Beans, bushels	5,696,433	8,585,405	3,514,221	6,263,209	9,147	213,381
Irish Potatoes, bushels	46,906,072	69,926,473	19,016,138	39,749,331	144,586	2,403,063

CHANGE OF NATIONAL EMPIRE.

Sweet Potatoes, bushels	19,834,295	21,301,242	19,432,803	20,488,724	214,860
Barley, bushels	4,216,763	6,169,419	5,258,605	5,258,605	4,457,867
Buckwheat, bushels	7,111,179	13,012,306	1,639,101	4,180,506	80,411
Value of orchard products	$5,268,494	$10,657,296	$2,435,701	$8,672,064	332	$1,262,615
Wine, gallons	73,861	356,645	89,333	1,081,187	18,971	249,360
Value of market garden prod	$3,805,046	$10,261,210	$1,285,580	$4,584,374	58,055	$1,273,904
Butter, pounds	200,161,124	257,882,960	111,968,204	167,217,182	189,384	4,572,630
Cheese, pounds	80,042,817	73,239,325	25,124,948	28,920,068	295,478	4,524,545
Hay, tons	10,213,308	10,826,040	3,308,028	6,822,187	68,128	359,769
Clover Seed, bushels	316,945	535,224	152,027	180,531	7,216	1,533
Grass Seed, bushels	300,226	282,184	116,583	563,227	6	4,629
Hops, pounds	3,291,549	10,148,740	205,422	272,014	22	1,162
Hemp, dew-rotted, tons	232	301	32,961	52,979	58	1
Hemp, water-rotted, tons	57	76	1,621	3,788		114
Hemp, other prepared, tons		8,574		13,680		
Flax, pounds	3,355,485	2,622,285	4,343,641	2,093,355	1,190	4,505
Flaxseed, bushels	213,140	142,130	349,047	424,668	5	69
Silk Cocoons, pounds	4,855	1,172	6,088	10,772		
Maple Sugar, pounds	22,577,904	28,050,706	11,675,532	11,989,699		
Cane Sugar, hhds 1000 lbs each	3,673	3,072	233,141	227,910		
Molasses, gallons, in 1850	760,030		11,940,879		82	
Cane Molasses, gallons, in 1860		1,005,000		13,968,396		46
Maple " "		464,378		1,277,770		26,342
Sorghum " "		575,872		6,056,909		1,327
Beeswax, pounds, 1860		539,638		721,827		18,353
Honey, pounds, 1860		10,359,970		12,541,339		
Beeswax and Honey in 1850	5,944,794		8,908,986		10	
Value of home made manufac's	$18,101,123	$8,414,632	$16,525,229	$15,739,436	$8,392	$402,788
Value of animals slaughtered	$40,799,400	$92,478,822	$30,564,074	$115,818,366		$4,433,444
Val. of agric'l impl'ts produced	$4,639,844	$8,903,815	$3,202,767	$8,883,594		$15,205
Val. of flour and meal produced	$88,151,308	$113,196,213	$45,857,566	$103,861,894	$1,888,332	$6,096,262
Val. of lumber sawed & planed	$37,134,449	$46,752,976	$19,037,922	$42,987,879	$2,349,005	$6,171,431
Val. of iron foundings	$15,340,012	$21,884,915	$4,771,505	$6,661,741		
Railroads, miles of	6,948	15,345	1,641	15,174		74
R. R., miles of, built in 10 yrs		8,307		13,533		74

In presenting it, we claim for it a superiority over any tabular statement of the material growth of the country that has ever appeared in public print. It contains upon its condensed surface the growth of centuries, and materials for volumes. At one glance the eye can scan the extent of territory, the population, the wealth, the industry, the live stock, the grains, the railroads, the progress, and the great working, moving embodiment of the country. Here, in one view, we can behold the growth of the most promising nation the world ever saw. Such is the progress exhibited that the growth of each ten years is equal to the growth of a nation. There is no parallel in history or experience for what we are, and none will ever surpass what we will be. Let us but labor to be as good as we will be great, and the solution of the problem of man's utility upon the earth will be solved before the close of another century.

By reference to the tabular statement, showing the material growth of the whole country from the Atlantic to the Pacific, it will be seen that in 1850 the States of the Atlantic slope were in advance of the Valley States in almost every practical and available interest belonging to the agricultural pursuits. Corn and wheat were the two principal products in which the Valley States excelled at that time. The Atlantic States had more land under cultivation, and a greater number of improved farms, the cash value of which was far greater than that of the Valley States; but the progress of ten years shows a wonderful change. When we compare the growth of 1850 with that of 1860, the advance is like the growth of a continent.

In 1850 the aggregate of improved lands in the Atlantic States was 63,965,491 acres, at a cash value of $1,991,599,378. In the Valley States the aggregate improved lands was 48,885,479 acres, at a cash value of $1,232,941,038. In 1860 the aggregate of improved lands in the Atlantic States was 73,882,853 acres, at a cash value of $3,132,561,500. In the Valley States the aggregate of improved lands was 87,034,199 acres, at a cash value of $3,446,702,533, showing, in the space of ten years, an advance of the Valley States over the Atlantic States of 13,151,346 acres of improved land, and a preponderance of cash value to the amount of $314,141,053.

In addition to this wonderful growth of the West, the States and Territories of the Pacific slope have advanced from 181,644 acres of improved land in 1850, at a cash value of $6,033,010, to 3,537,668 acres of improved land in 1860, at a cash value of $67,780,934. These figures are most gratifying in their showing. The whole growth of the West, in agricultural pursuits, is unparalleled in the history of the human race, and yet the Republic is in its infancy. Massachusetts has but little more than one-half her acres under cultivation, while Illinois has far less than one-half her lands in farms. The improvements of the other States, in all the kindred elements of agriculture, are about in the same ratio. But what are these half developments when compared with the full growth of the country? The territory of the Valley States is more than three times as large as that of the Atlantic States, and, with its incomparable advantages for agriculture, must lead the way in the pursuit of husbandry.

We must comprehend that with the growth of the Republic must be the intellectual and moral growth of the people. As the nation expands, so must the legislative and moral mind expand to comprehend its demands and necessities. The legislator must comprehend that the laws are yet to be of broader significance, and the moralist and the educationalist must also learn that precept and discipline must extend beyond to broader fields of use than heretofore; and may we not hope that, at no distant day, some genius may arise who will add to the material statistics of the country the statistical growth of the morals and intellectual advancement of our people, and thus furnish the measure of our most valued growth?

DEMAND FOR A CHANGE OF SEAT OF GOVERNMENT

—AND ITS—

LOCATION AT ST. LOUIS.

Enlightened public sentiment everywhere demands the appropriate use of all public interests. There can be but two essential considerations enter into the subject of locating the seat of government in any nation. One is the propriety of locating it where it can be easily defended in time of war; the other of locating where it will best subserve the public and special interests of the people of the government. The history of nations furnishes no considerations greater than these. It is probably true that the greater number of national capitals of antiquity became fixed by reason of the prestige of civil and commercial power being invested in certain places at the time of revolution or governmental changes. Some nations have considered the subject of locating their seats of government in a secure part of the country, but most capitals have been located where they would best accommodate the commercial interests of the people and the business interests of the industrial masses. In proof of this statement we have but to trace the map of the history of mankind, and almost everywhere we see governments yielding to the sway of commerce, whether upon the sea-coast or upon inland waters. While it is true that the argument in favor of locating so as to protect them from invasion is one of barbarian origin, it is still of some consideration to mankind, and cannot be heedlessly overlooked by this people. This consideration, as I have already stated, had some weight with the first Congress when the subject of a permanent seat of government was discussed. But the one which was of greater concern to them was the one which this people cannot overlook at the present time, and that was the argument in favor of locating the seat of government centrally, so as to accommodate the business interests of the American people. This is the argument that I shall endeavor to set forth in this pamphlet in its broadest and most significant form.

No people in the world are so directly and so much interested in government as the American people, for theirs is "a government of the people, by the people, and for the people;" and to-day Washington city, with the seat of government conventionally there, is not a practical reality to the great majority of the free, enlightened, and industrious American people. It is an effete thing of the past; it is a place of no great interest in common with the great wealth, industry, and progress of the American nation. It is a tomb "full of dead men's bones," having neither beauty within nor without. It has served the purposes of the Old Government, and held the nation in bondage, and now it is unfit for the purposes of the New Republic. It is henceforth a disgrace to the Republic that, with its 40,000,000 inhabitants and its 40,000 miles of railway, its Capital should remain in a city of 100,000 inhabitants, and only one railroad going to and passing through it, and that, too, the most anti-American monopoly on the continent.

There is an instinctive feeling pervading the American people that the growth of the Republic has rendered Washington unfit to remain its Capital, and that, in subserviency to the will and demands of the people, the seat of government will be moved at an early date to the great Mississippi Valley—to the banks of the Father of Waters—to St. Louis, occupying as she does substantially the geographical center of the nation. Especially has she the most important points of special interest favorable for the seat of government, as I shall still further point out. No place in the nation is more suitable for the seat of government. Every American writer who has spoken at all in public print has pointed to St. Louis as "the future home of the seat of government." So also does the American statesman look to St. Louis as the most favored place for the Capital of the New Republic; and true to the instinct of the people will it come, and that, too, before another five years passes away. This change of national empire is of more vast concern to the future existence and welfare of the Republic than most people think. The Atlantic slope and Washington City are but the cradle and place of national life in our governmental infancy, and it is intolerable and impossible to attempt the continued confinement of our national life to that cradle after the nation in its maturer years has grown far away and far too great for that place of childhood. Nothing is more deceptive than to think that this great people will not make the change.

THE GEOGRAPHICAL ARGUMENT.

St. Louis is situated on the west bank of the Mississippi river, 1,350 miles from its mouth, and about 1,100 miles from the Lake of the Woods on our northern boundary. It is about 1,000 miles west from New York, and 2,000 miles east of San Francisco. In its relation to our northern and southern boundaries it occupies substantially the geographical center of the country. Its position, when considered from the East and West, is not central, geographically speaking; yet I will show by the population, commercial, political, and conclusive arguments, that its geographical position, in reference to the East and West, is adjusted, and thus rendered the favored place for the seat of national empire for the New Republic. Even were there no other argument, it would be sufficient in itself to show that by far the greater portion of our national domain lies beyond the Mississippi river, and that the future unfoldment of the nation will be there.

At least 10,000 miles of navigable rivers bear their commerce in the interest of St. Louis. And such is its geographical position that it must be the vitalizing heart of the wealth, the industry, the civilization, the politics, and the social progress of the Mississippi Valley. No inland place on the continent holds so favored a position. It is the great point of radiation. Situated in the vicinity of the mouths of the Ohio, the Missouri, and the Illinois rivers, with New Orleans on the south, Chicago on the north, New York on the east, and San Francisco on the west, St. Louis cannot fail to hold the most important position of any city on the continent.

But there is still a higher sense in which to view this matter. It will be found, by a close examination of the career of mankind upon the earth, that they have lived and journeyed around the earth in the what has been called an isothermal zodiac, or belt of equal temperature, which girdles the earth in the north temperate zone. Within this girdle or zone are all the civilized nations of Asia, Europe, and America, and about 850,000,000 or nine-tenths of the human race. Within this isothermal or human zodiac or zone is an axis indicating the central or life line of this zodiac around the globe. Starting from the Orient, in Hindostan, near Bombay, it passes northward through Persia, Arabia, the Mediterranean, France, England, thence to New York, Pittsburg, St. Louis, and on to the Pacific Ocean.

"It is along this axis of the isothermal temperate zone of the northern hemisphere that revealed civilization makes the circuit of the globe. Here the continents expand—the oceans contract. This zone contains the zodiac of empires. Along its axis, at distances scarcely varying one hundred leagues, appear the great cities of the world, from Pekin, in China, to St. Louis, in America. During antiquity this zodiac was narrow; it never expanded beyond the north African shore, nor beyond the Bontic Sea, the Danube, and the Rhine. Along this narrow belt civilization planted its system, from oriental Asia to the western extremity of Europe, with more or less perfect development. Modern times have recently seen it widen to embrace the region of the Baltic Sea. In America it starts with the broad front from Cuba to Hudson's Bay. As in all previous times, it advances along a line central to these extremes in the densest form and with the greatest celerity. Here are the chief cities of intelligence and power—the greatest intensity of energy and progress.

"Science has recently very perfectly established by observation this axis of the isothermal temperate zone. It reveals to the world this shining fact, that along it civilization has traveled, as by an inevitable instinct of nature, since creation's dawn. From this line has radiated intelligence of mind to the North and to the South, and towards it all people have struggled to converge. Thus, in harmony with the supreme order of nature, is the mind of man instinctively adjusted to the revolutions of the sun and tempered by its heat."

No relation of man to mother earth is more interesting than this fact of his essential alliance to this zone by instinct, as it were, and thus guided on his journey around the planet.

When we trace the axis of this zone, as thousands of years of history have indicated it by temperature and population, encircling the earth like a great magnetic chord, we conceive that when it passes over our continent, like the electric wire that is hung in the fork of the tree, so does this great axis, in passing our continent, find lodgment in the forks of our great rivers, thus passing within the shadow of our city, and, as with an enchanter's wand, by its touch awakes St. Louis to an imperial greatness and destiny.

THE POPULATION ARGUMENT.

The population argument is one of the most interesting features of this subject, and the one upon which everything else depends. In 1790, or about the time of the adoption of the Federal Constitution, we had a population of 3,929,827, which was a little more than the present population of the State of New York. But we have grown up since then, and within the lifetime of a human being, to a population of 31,443,322, and will at the census of 1870 be increased to more than 40,000,000. The population of the United States in 1850 was 23,191,876, which in ten years, or by the time of the taking of the census in 1860, had increased 35.52 per cent.

The increase of the population of the Northwest during the last ten years has been 67.9 per cent., while the ratio of increase in the whole country has been 35.52. The population of the Northwest, by the census of 1860, was 28.85 per cent. nearly one-third. Of the total increase in the population of the country, 44.67 per cent. was in the Northwest alone. An increase at the same ratio during the present decade will give the Northwest in 1870 a population of 15,212,622—an increase of 6,139,567. Massachusetts, the most densely populated of all the States, has 157.8 inhabitants to the square mile. A like density of population in the Northwest would give us a population of 133,011,198. A density of population equal to that of England (230 per square mile) would give an enumeration of 279,846,120.

The popular vote of 1852 is copied from the census compendium (1850), p. 50; that of 1860, from the census returns. Under the old apportionment (1850), the Northwest had 24.31 per cent. of the members of the House of Representatives, or a fraction less than one-fourth. Under the census of 1860, she is entitled to 30.47 per cent., or nearly one-third. At the Presidential election of 1852, the Northwest cast 29.46 per cent. of the popular vote. In the Presidential election of 1860, she cast 36.24 per cent. of the popular vote—more than one-third. In the electoral college in 1860, the Northwest cast 23.14 per cent. of the vote for President and Vice President.

The following table shows the standing of the *loyal* States in respect to political power in 1852 and 1860:

	1852.	1860.
Popular vote for President	2,583,918	3,805,640
Electoral votes	205	
Under the new census		210

In 1852, the Northwest cast 35.68 per cent. of the popular vote for President in the loyal States, and 34.63 per cent. of the electoral vote. In 1860, she cast 44.4 per cent of the popular vote, and in 1864 had 40.63 per cent. of the votes of the loyal States in the electoral college.

In England, the density of population is about 230 persons to the square mile; but England is in some measure the workshop of the world, and supports by her foreign trade a greater population than her soil can nourish. In France, the density of population is about 160 to the square mile. In Germany, it varies from 100 to 200.

Assuming, on these grounds, that the number of persons which a square mile can properly sustain in our rich country, without generating the presence of a redundant population, is 490 (the number authorized by a writer in the *Britannica Encyclopedia*), this would, when the country is fully developed, give to the Atlantic slope a population of 219,970,310, and to the Valley States a population of 761,302,530, and to the Pacific slope a population of 483,754,460, and to the whole country a total population of 1,465,027,400—a body of people infinitely beyond the comprehension of the human mind. Even the half of this number of inhabitants would make us the greatest nation that ever ruled on earth.

The estimate above gives us a population greater than the entire present population of the world. But the grandeur of the thought still swells when we consider that in a little more than a century, or beginning with a new era, our numbers will well nigh approximate this great growth.

The extraordinary tendency of population to this country from the over-populous regions of Europe overrides all theories of a systematic increase of our population for the present time. Yet another census will give the Mississippi Valley a preponderance of population over the Atlantic slope, and more than double that of the Pacific slope in ten years. But from the extraordinary increase of population which our growing country has sustained, we find ample hope for our population to reach that of multiplied millions, by the systematic and philosophical theory of Malthus, the great father of political economy, which no credible writer has controverted. He laid down as a law for human increase that the productive powers of healthy, well-fed, well-lodged, and well-clothed human beings was naturally so great that fully two children will be born for every person who will die within a given time; and George Combe, commenting upon this theory of Malthus, said that population would double itself every twenty-five years. He added that this increase took place in the new States of North America, independent of immigration. Then, taking the Malthusian doctrine for our guide, we would have at least 100,000,000 inhabitants in the United States at the close of this century. But, with the aid of immigration, that number will be reached before the close of the century.

Contemplating this vast increase, with its density, in the Mississippi Valley, who is so blind as not to be able to see that it is the right of the wealthy and powerful to possess the seat of government? This subject of population throws the argument more especially in favor of St. Louis when we consider that the distribution of population will not be uniform over the country, but will be much denser along the rivers and lakes where commercial, agricultural, and mechanical pursuits go hand in hand. The history of mankind, all over the earth, shows the dense population to be gathered along the maritime shores. Look to the Babylonians, the Tyreans and Sidonians, the Carthagenians, the Levantes, the Romans. Look to modern Europe—England and France. Such will be the truth in our own land. Between the east line of Kansas and the Rocky Mountains will never be a dense population, for the country is not adapted to a variety of industrial pursuits, such as can be prosecuted with profit. That portion of our domain will be the great American pasture, and will be devoted to stock-growing as the chief pursuit, and therefore will not, cannot, be densely populated. This will give the preponderance of population to the Mississippi river and her tributaries, and counterbalance the disproportion of geographical position, and more than make up with population for St. Louis what she loses by geography, thereby rendering her position the most favorable.

Illinois and Missouri will not only soon become the two great and powerful States of the Mississippi Valley, but also of the nation; and with their rich soils, their valuable minerals, their timbers, their water-powers and navigable advantages, in short their supreme advantages for all the industrial pursuits of civilized life, they are destined to support a greater number of American citizens to the square mile than any other States of the Union; and St. Louis will, in her future growth, extend her limits to Alton, Collinsville, Belleville, Jefferson Barracks, Kirkwood, and, in another century, be a city of 10,000,000 inhabitants. Such is American progress, and such is American destiny.

The future reveals nothing but greatness — nothing but that onward progress and greatness which is everywhere seen and felt to be approaching. Let us all anticipate it and be energized by the thought that multiplied millions of wiser and better people than we are soon to take our places and move on in the great caravan of life, as Divine law will direct.

THE COMMERCIAL ARGUMENT.

There can be no mistake about St. Louis occupying the most favorable commercial position of any inland city in the Mississippi Valley. It was no mere fancy of Pierre Laclede Liguest that caused him to select this favorable position for a great city; in fact, there seemed to be a kind of instinct that pointed the early French pioneers to the most favorable town sites in the great West. At the time when the seat of government was located at its present place, it will be remembered that where St. Louis now stands did not belong to the United States, nor did a single foot of land west of the Mississippi river, and at that time St. Louis was only a trading post or village of about one thousand inhabitants. It was founded on the 15th day of February, 1764.

The first item of importance to St. Louis, as a great commercial city, is its location, standing as it does on the great Mississippi river, upon whose waters now does and forever will float the greatest inland commerce in the world. It commands the trade of 10,000 miles of the most valuable river navigation on the continent, and is the only city of importance on the Western waters where steamboats come to discharge their freights and reload and return. It is essentially a distributing port. No boats of any value pass its harbor. To its 10,000 miles of river navigation let us add 10,000 miles of railway communication; then let us go forward but a year or two to that commercial triumph of the West, when her trade, true to a law of nature, will follow the great waters of the gulf as surely as the waters themselves find their way there; then with the 10,000 miles of river navigation and 10,000 miles of railway communication, and with these rivers and railways bringing a rich commerce from other lands and from all over the continent, who will not look with delight to St. Louis as destined to be the great city of the Mississippi Valley—the great inland depot of the continent?

Touching the commercial importance of St. Louis, it is well that we look beyond and see what shape the commerce and industry of the country will take.

At this time there is a continental strife for commercial supremacy inaugurated between the Atlantic cities and the people of the West. The contest is for the purpose of determining whether the trade of the West shall go across the continent to the Atlantic cities, or whether it will go down the Mississippi river and her tributaries to the Gulf, and from thence to the markets of the world. In this contest the West will triumph and her products follow the water courses. The question will be settled in the next three years.

Following this contest will come that long-anticipated change, or at least the time for it, when a railway is completed to the Pacific Ocean, and we look for our trade with China and India to find its way to us through different channels. In this matter the people have no doubt over-estimated the importance and magnitude of that great continental change in our foreign commerce. It is true the completion of those great railways will be a wonderful triumph of American industry; but their completion will not bring such a change and such an era in our continental development as many have anticipated. On the other hand, the great commercial and civil era to which we are approaching will come, with our industrial and commercial tendency, to the tropics of our own hemisphere. In industry the destiny of this people is a continental conquest. Nothing but wild and foolish extravagance and impracticability will lead our people over distant oceans to distant lands for products, when we have at home all the climates, all the soils, and all the advantages that the globe can afford. Nor will the American people act so foolishly. It is not in their experience to do so. They will do otherwise. Already there is a great trade in the tropics, which our people can easily command if they do but make the proper use of the means within their reach.

The following remarks of Judge Burwell, of New Orleans, taken from a speech of his before the St. Louis Board of Trade,

Monday, October 19, 1868, are full instruction upon the trade of some ports of the tropics, and should have great weight in influencing the people of the West in their commercial action:

A COMMERCIAL POLICY FOR THE AMERICAN CONTINENT.

But those who will examine the present trade relations between the United States and Territories referred to will find very many expensive and vexatious impediments and charges that should be and may be removed by a proper exercise of diplomatic influence.

It is impossible to part from this subject without calling to your attention the importance of the Cuban trade and the singular facilities which exist for securing it to the United States. It is not the province of a commercial discourse to provide for or even propose the acquisition of the Island of Cuba, but it may be reasonably expected that all the impediments to a direct commercial intercourse will be removed.

There is no reason why a reciprocity treaty may not as well be made in regard to Cuba as Canada. But supposing all trade impediments removed, how attractive the commercial prospect? Cuba produced last year $259,000,000 value; of this, perhaps, thirty millions dollars in sugar, coffee, and other products have been imported into the Western States and Territories. Cuba consumed, perhaps, 500,000 barrels of foreign flour, besides other provisions, and this could be supplied by the Western States and Territories. Already Havana is within less than one hundred hours of St. Louis and Chicago by steamer and rail. This time could be reduced considerably. But with the removal of all impediments to a free interchange of commodities between these great and reciprocating interests, how extensive and how precious must be the commerce. A similar estimate may be made in regard to the trade between New Orleans, Vera Cruz, and other Gulf ports. The Isthmus of Panama, however, presents the most attractive prospect of gathering an immediate harvest. Last year there crossed the Panama Railroad more than one hundred and fifty thousand tons of freight, with some thirty thousand passengers and perhaps thirty millions of the precious metals. While much of these last items will be, of course, diverted to the Pacific railroad, there will be always an important value of commerce crossing at that point. Take, then, what we will estimate the proportion due to the Valley of the Mississippi and of the lakes, it will be seen that this portion can be readily taken direct from Panama to Chicago and St. Louis by way of New Orleans. The distance from Panama to New York is 2,300 miles, or ten days of gulf and ocean voyage, with its proportionate insurance. The distance from Panama to New Orleans is 1,700 miles, or about seven and a half days, with insurance in proportion. From New York to Chicago is about 950 miles rail, to St. Louis is about 1,200 miles all rail; from New Orleans to Chicago, all rail, will be about 950 miles, and from New Orleans to St. Louis something less. These are approximate and not exact distances; but they show that a passenger or a package at Panama, destined for either Chicago or St. Louis, could reach Chicago in some three days' less time by way of New Orleans than by way of New York, and would reach St. Louis with still less time and distance. These sketches are but suggestive, for it is impossible to go into the details of a subject so extensive.

But do they not all point to the importance of an organization of the interests concerned in this great commerce? Do they not justify these interests in making a combined demand on Congress to view with equal favor the inboard line of transportation and its outlet, as the coast line of the Atlantic or Pacific? If Congress gives aid to steam lines from New York to Rio, or to Vera Cruz, or to Havana, or Panama, or from San Francisco to China, should it not, in common justice, aid steam lines from New Orleans to the Gulf and Atlantic ports, and even to those of Europe?

But to organize this trade will require certain combinations between the river and ocean steamers. A first-class ocean steamer for the Rio trade will cost $100,000. She will carry out nine thousand five hundred barrels, and bring back fifteen thousand sacks of coffee. The voyage, out and in, from New Orleans may occupy about seventy-five days — equal to about five trips a year. Now, if the lake and river cities will take joint stock in such a line of steamers, and run their railroads, steamboats and barges in close connection with them — prorating for distances, signing through, and consigning to each other without other than actual charges, it is perfectly plain that each line must load the other, and that the whole freights thus apportioned will support these lines as they do those which now conduct it. This enterprise is equally as applicable to the communication between Chicago and St. Louis, and Havana, Vera Cruz and Panama, as with Rio. Develop a continental market for American products. To impress on a commercial audience the immense importance of requiring Western members of Congress, without respect to politics, to demand of the existing and future administrations the removal of these obstacles to our trade, it may be proper to remind them of its extreme value. The general trade of the continent and islands south of the United States was estimated ten years ago at $3500,000,000. It comprises many products not cultivated elsewhere than at or near the tropics. It brings to Western millions products desired by civilized man. It affords woods for use and ornament; drugs for our climatic diseases. In return, this market demands a very large supply of Western provisions. Now, when we regard the immense development of European productions by the same means of artificial transportation with our own—when we note the transfer of a productive force by immigration from Europe to America, it is obvious that the grain and other productions of the Ukraine, the Don, and the Danube, will be poured in increased volume into the consuming markets of Europe. This form of consumption will react, no doubt, upon the abundant provision supply, and perhaps, except in seasons of famine or low wages, consume about all that can be imported from other countries. But the United States lies near the whole of our southern continent, and it can deliver its provision crops, with proper facilities, at a cheaper rate than the European farmer, who must cross four thousand miles of intervening ocean. Why, then, should not some statesman of the West, emulating the example and perpetuating the ideas of your own great Benton, take up this subject and consummate the great exchange of Western provisions for tropical products. It is a work worthy the ambition of a patriot, and the prosperity which would follow would raise his renown above the grade of military glory or the successful diplomacy of the most astute politician. Let me, then, give an example of the trade which might be advantageously opened with Brazil, Cuba, and Mexico. Congress has committed the guardianship of commerce and finance to the city of New York, or rather that great city has secured, by her energy and enterprise, the stewardship of the Union. She has now postal subventions to Rio, Havana, Panama, Vera Cruz, as the Pacific city of San Francisco has to China. The Government now pays a line of steamers, running between New York and Rio, $150,000 for twelve round trips per annum. Every practical merchant will see what immense aid this must afford this line in any contest for the Rio trade. Now, if Congress will divide this subvention, or give as much to a similar line between New Orleans and Rio, or, indeed, suspend any appropriation to either, it is very obvious that a line from New Orleans, running in connection with our river vessels, must possess great advantages, as far as St. Louis and Chicago are concerned. It is obvious that the voyage between Rio and New Orleans is not longer than between New York and Rio. Now, a cargo of flour being at St. Louis can surely be shipped at less cost by the river to New Orleans than by rail to New York; and a cargo of coffee imported can undoubtedly be brought cheaper in return. For the first time, perhaps, this great inboard line of river and rail communication between the South and North is prepared to compete with the coastwise route. Will it not promote the interests of the West to pay its freights for this transportation to its own railroad, steamboat and barge companies, rather than to coastwise shipping and coast-route lines owned and operated elsewhere?

The people of St. Louis and the West must learn that next in importance to the Mississippi river is a railway through the Southwest to Galveston, thus making a great trunk line from Chicago *via* St. Louis to the Gulf, and uniting the Gulf and the lakes at a distance of about 1,000 miles, and St. Louis to the Gulf at a distance of about 700 miles. Akin to this road in importance will be another from Denver City into Mexico. By these means will be won a commerce from the tropics and South America surpassing the distant trade of the Orient. With all these future developments of our continental and foreign trade, St. Louis will still remain the central city and commercial depot of the country; and with the minerals and coal of Missouri and Illinois, the timber and the water, the great workshops of the country will be hers.

The following article upon South American commerce, from the *American Gazette*, published at Philadelphia, is also interesting to the subject:

In the last five years the tonnage of the United States shipping has fallen from 6,000,000 tons to 3,360,000. In the same period the foreign shipping trade to the United States has increased from 2,600,000 to 4,500,000 tons. The increase has been mainly British. The decrease has been exclusively American. We do not cite the facts as a discouragement, but as an incentive. They show, if anything can show, how important it is that effectual measures should be immediately undertaken to restore to us that prosperity in this important field that once belonged to us.

Foremost among the opportunities of the moment for creating this restoration is the opening of the navigation of the Amazon to the flags of all nations. In this estimate we by no means overlook the advantages immediately and prospectively accruing from closer commercial relations with China, Japan, and Western South America, and from the increased whaling, fishing, and lumber business of the Northern Pacific. They have their several places, and very important they are too. But the facilities for business given by the opening of the Amazon may, if immediately and wisely improved, go as far as any in enlarging our mercantile marine until it swells beyond its furthest limits. The Amazon was opened by Brazil on the 7th of September, and all of its tributaries — the Tocantins and the San Francisco — are free for direct trade to all countries. Now Brazil is one of the largest empires in the world. It is twelve times as large as France, and comprehends a surface of 2,700,000 square miles, with a seacoast of more than 4,000 miles; borders on nearly all the States of South America, as well as on the British, French, and Dutch possessions, and its river system is equal and in some respects superior to that of this country. The country is thinly populated, and only about half civilized in the interior, from which it may be calculated how great a field for trade is opened, and how rapidly civilization and business will follow a wise, liberal, and energetic cultivation of our commercial relations there. We may treble our trade in three years.

For hundreds of years the Amazon, navigable for nearly four thousand miles through the heart of Brazil, has been closed to foreigners. Fewer vessels pass between any two points on it in a year than run from St. Louis to New Orleans in a month. The tides of the Atlantic, into which it flows through an embouchure 186 miles across, are felt 400 miles from its mouth, where the water is twenty fathoms deep and the river more than a mile wide. Its banks and the interior on either side produce maize, rice, coffee, sugar, cotton, tobacco, spices, timber, medicinal plants, cattle, gold, iron, and lead. Bolivia opened the tributaries to the Amazon from her territory to all countries in 1853. Brazil neutralized the uses of this concession by closing the Amazon, through which alone the Bolivian streams could be reached. Now, however, this is changed, and the Bolivian concession may be improved. The Tocantins, a tributary to the Amazon, is 1,200 miles long. It threads very fertile countries, but is not continuously navigable, owing to falls. It is opened by decree, however, to Cameta, with 40,000 inhabitants, and from Madeira to Manaos, provinces rather than cities. The San Francisco, the other river recently opened, is about 1,300 miles long, but, owing to obstructions, cannot be navigated any higher than Penedo. At intervals it is navigable beyond for 200 miles together. The current will carry vessels 100 miles in twenty-four hours. Gold is among its deposits. Nitrate of soda is found on its banks; and in one spot, a valley sixteen leagues broad and twenty long, it is found on the surface, and everywhere is procurable with little labor. The article is so valuable as a fertilizer that our demand for it might alone maintain a large commerce.

Whatever advantages are to be gained from the opening of these rivers should inure to our commerce. The winds and currents are so much in our favor that a chip thrown into the Atlantic at the mouth of the Amazon would float by Hatteras, in the very line of navigation; and our Atlantic ports of Baltimore, Philadelphia, and New York, are natural half-way houses between Para and Europe. What we need is the prudent enterprise to improve this opening. Had it occurred in 1858, or at any subsequent day to the rebellion, our flag would already have been there, and our exchanges would have been made along the whole banks of the streams. But now we hesitate. Thus far we are aware of nothing that has been done. As such advantages cannot long lie idle, we look to an early moment when they will be seized, and another aid given thus to that renewed maritime activity that should date from this year. If the opportunities afforded in Eastern Asia, in the Northern Pacific, on both coasts of South America, and now into the very heart of that part of the hemisphere, are not mirrored in a rapid commercial improvement, we shall begin to believe that the nature of our people has changed, and that they are at last unable to see or unwilling to improve occasions of the greatest value.

These two preceding statements show to the American people a field of commerce more inviting, at our own doors, than can be found far away in the Orient; and every consideration of our industry and commerce demands that no opportunity be lost in establishing the most liberal governmental policy by which to secure it.

By cultivating that rich trade of the Southern countries, the Atlantic sea-board cities, and those of the Gulf, will gain back more in value than they will lose of the Orient trade after the completion of the Pacific Railway.

THE POLITICAL ARGUMENT.

There is still another way by which we can demonstrate the growth and preponderance of power in the West over the old federal arrangement of the Government. It is by showing the approaching supremacy of political power in the union of the Valley States with those of the Pacific slope. The Atlantic slope has an area of 423,197 square miles, which is divided into seventeen States. Under the Constitution they are allowed 34 Senators and 120 Representatives in the National Legislature. The Mississippi Valley has an area of 1,899,811 square miles, with less than one-third of its territory made into States. It now has eighteen States, which under the Federal Constitution are allowed 36 Senators and 115 Representatives in the National Legislature. The Pacific slope has an area of 627,256 square miles, part of which is made into three States, which are entitled to six Senators and five Representatives in the National Legislature. Alaska has an area of 577,390 square miles, and is large enough to make more than fourteen States as large as Ohio. Another view of our country shows 860,000 square miles east of the Mississippi river, which is already divided into twenty-seven States, including Louisiana and West Virginia. These send 54 Senators and 205 Representatives to the National Legislature. West of the Mississippi river we have 2,070,000 square miles, exclusive of Alaska, which at the least calculation ought to be made into fifty new States, each one of them being larger than Ohio and containing 40,000 square miles.

By these figures it is easy to be seen that the great preponderance of political power will soon be far removed from the Atlantic slope.

It is safe to say that the census returns of 1870 will show the Mississippi Valley to more than double the Atlantic slope in many departments of the national wealth of the country, and therefore compel the preponderance of taxation west of the Alleghany mountains. Already, by the census of 1860, the cash value of farms in the Mississippi Valley, the value of farming implements, the value of live stock, and the value of animals slaughtered, was far in advance of the Atlantic slope. Not only will the greater portion of the taxes to support the Government come from the West, but also will the popular vote of the country be far greater in the Mississippi Valley; and thus in every way the political power of the Republic will soon be found far away from the present seat of government. Nothing is more certain than this; and who is so foolish as not to be able to comprehend that political power will not, when backed by the wealth, the genius, and the preponderating millions of American citizens, demand that the seat of government shall come to the Mississippi Valley? Even now the preponderance of power is in the hands of the West, and her people have made up their minds to ask for its removal before another Presidential term expires.

Aside from numerical numbers and territorial extent, the cohesive power of nationality demands that the ruling power of a nation be located in the midst of its material power. The life of a nation is made doubly secure when united with the strongest and greatest commercial and material interests of its people; for thus united they become a complement in purpose and destiny — the security and perpetuity of the one becomes the security for the perpetuity of the other. Philosophy is alike applicable in the institutions of men as in the works of nature, and nothing can be more absurd than to imagine that the life and perpetuity of this Republic is as secure for the future, with the seat of government at Washington — a distant place on the outskirts of the country, with no material power or commercial prestige — as it would be at a central position in the Mississippi Valley, where the great vitalizing heart of the Republic beats in keeping with its onward march of progress and greatness.

THE CONCLUSIVE ARGUMENT.

Perhaps the reader of this little pamphlet, if his mind was not already favorable to its cause, will be satisfied of its justness before he reaches this conclusive argument. If not, I only ask his further consideration of the matter, and then demand of him an impartial decision.

The statement of the Old Government and the map show that the location of the Capital of the Nation by the first Congress, in 1790, was an act totally supported by local and incidental circumstances, belonging wholly to that period, and not of any value whatever at the present time.

The map and statement of the New Republic show, in the most clear and conclusive manner, the existence of entire different incidents and circumstances at the present time, and that the circumstances and incidents demand a change of national empire, or the removal of the seat of government to the Valley of the Mississippi.

I have shown the wonderful growth of the nation, in territorial extent and material power, since the adoption of the Federal Constitution, to be essentially beyond the limits of the thirteen States of the Old Government, and far away to the Mississippi River and the Pacific Ocean. I have endeavored to show that that growth, with its transfer of material power to the Valley States, creates, by means of our geographical expansion, our immense increase of population, our vast internal and Western commerce, and our political power, a demand for the removal of the seat of government to the Mississippi Valley.

But, in a further and more conclusive consideration of the subject, the attention of the reader is asked to more evidence of the material growth and vastness of the Republic. To begin the submission of that evidence, the following table is offered as a basis:

Historical and statistical table of the United States of North America.

[NOTE.—The whole area of the United States, including water surface of lakes and rivers, is nearly equal to four million square miles, embracing the Russian purchase.]

The thirteen original States.	Area in square miles.	*Population, 1860.
New Hampshire	9,280	326,073
Massachusetts	7,800	1,231,066
Rhode Island	1,306	174,620
Connecticut	4,750	460,147
New York	47,000	3,880,735
New Jersey	8,320	672,035
Pennsylvania	46,000	2,906,115
Delaware	2,120	112,216
Maryland	11,124	687,049
Virginia—East and West	61,352	1,596,318
North Carolina	50,704	992,622
South Carolina	34,000	703,708
Georgia	58,000	1,057,286

States admitted.	Act Organizing Territory.	U. S. Statutes.		Act admitting State.	U. S. Statutes.		†Area in square miles.	*Population—1865.
		Vol.	Page.		Vol.	Page.		
Kentucky				Feb. 4, 1791	1	189	37,680	1,155,684
Vermont				Feb. 18, 1791	1	191	*10,212	315,098
Tennessee				June 1, 1796	1	491	45,600	1,109,801
Ohio	Ord'ce of 1787			April 30, 1802	2	173	39,964	2,339,502
Louisiana	March 3, 1805	2	331	April 8, 1812	2	701	*41,346	708,002
Indiana	May 7, 1800	2	58	Dec. 11, 1816	3	399	33,809	1,350,428
Mississippi	April 7, 1798	1	549	Dec. 10, 1817	3	472	47,156	791,305
Illinois	Feb. 3, 1809	2	514	Dec. 3, 1818	3	536	*55,410	1,711,951
Alabama	March 3, 1817	3	371	Dec. 14, 1819	3	608	50,722	964,201
Maine				March 3, 1820	3	544	*35,000	628,279
Missouri	June 4, 1812	2	743	March 2, 1821	3	645	*65,350	1,182,012
Arkansas	March 2, 1819	3	493	June 15, 1836	5	50	52,198	435,450
Michigan	Jan. 11, 1805	2	309	Jan. 26, 1837	5	144	*36,451	749,113
Florida	March 30, 1822	3	654	March 3, 1845	5	742	59,268	140,425
Iowa	June 12, 1838	5	235	do	5	742	55,045	674,948
Texas				Dec. 29, 1845	9	108	*274,356	604,215
Wisconsin	April 20, 1836	5	10	March 3, 1847	9	178	53,924	775,881
California				Sept. 9, 1850	9	452	*188,981	305,439
Minnesota	March 3, 1849	9	403	Feb. 26, 1857	11	166	83,531	173,855
Oregon	Aug. 14, 1848	9	323	Feb. 14, 1859	11	383	95,274	52,465
Kansas	May 30, 1854	10	277	Jan. 29, 1861	12	126	81,318	107,206
West Virginia				Dec. 31, 1862	12	633	23,000	
Nevada	March 2, 1861	12	209	March 21, 1864	13	30	†112,090	86,857
								10,507
Colorado	Feb. 28, 1861	12	172		13	32	*104,500	34,277
								2,261
Nebraska	May 30, 1854	10	277	March 1, 1837	13	47	75,995	28,841

CHANGE OF NATIONAL EMPIRE. 65

Territories.	Acts organizing Territories.	U. S. Statutes. Vol.	U. S. Statutes. Page.	Area in square miles.	*Population.
New Mexico	Sept. 9, 1850	9	446	121,201	The estimated population of these Territories on Jan. 1st, 1865, as above indicated, was 360,000.
Utah	......do......	9	453	†88,056	
Washington	March 2, 1853	10	172	69,994	
Dakota	March 2, 1861	12	239	240,597	
Arizona	Feb. 24, 1863	12	664	**113,916	
Idaho	March 3, 1863	12	808	90,932	
Montana	May 26, 1864	13	85	143,776	
Indian Territory				68,991	
District of Columbia	{ July 16, 1790 { March 3, 1791	1 1	130 214	{ 10 miles { square	
***N. Western America, purchased by treaty of May 28, 1867				577,390	70,000

* The total population of the United States in 1860 was, in round numbers, 31,500,000. In 1865 it is estimated that the population was 35,500,000, including the inhabitants of the Territories, estimated at 360,000 persons on January 1, 1865. At the present time, November 1, 1867, according to the most satisfactory estimate, it is about 38,500,000. In 1870, according to existing ratios, the population of this country will be over 42,250,000. At the end of the present century, 107,000,000.
† The area of those States marked with a star is derived from geographical authorities, the public surveys not having been completely extended over them.
‡ The present area of Nevada is 112,090 square miles, enlarged by adding one degree of longitude lying between the 37th and 42d degrees of north latitude, which was detached from the west part of Utah and also northwestern part of Arizona Territory, per act of Congress, approved May 5, 1866; U. S. Laws 1865 and 1866, page 43, and as assented to by the legislature of the State of Nevada, January 18, 1867.
§ White persons.
‖ Indians.
¶ The present area of Utah is 88,056 square miles, reduced from the former area of 106,382 square miles by incorporating one degree of longitude on the west side, between the 37th and 42d degrees of north latitude, with the State of Nevada, per act of Congress, approved May 5, 1866, and as accepted by the legislature of Nevada, Jan. 18, 1867.
** The present area of Arizona is 113,916 square miles, reduced from the former area of 126,141 square miles by an act of Congress, approved May 5, 1866, detaching from the northwestern part of Arizona a tract of land equal to 12,225 square miles, and adding it to the State of Nevada. U. S. Laws 1865 and 1866, page 43.
NEVADA.—Enabling act approved March 21, 1864; Statutes, volume 13, page 30. Duly admitted into the Union. President's proclamation No. 22, dated October 31, 1864. Statutes, volume 13, page 749.
COLORADO.—Enabling act approved March 21, 1864; Statutes, volume 13, page 32. Not yet admitted
NEBRASKA.—Enabling act approved April 19, 1864; Statutes, volume 13, page 47. Duly admitted into the Union. See President's proclamation No. 9, dated March 1, 1867. U. S. Laws 1866 and 1867, page 4.
That portion of the District of Columbia south of the Potomac river was retroceded to Virginia July 9, 1846. Statutes, volume 9, page 35.
*** BOUNDARIES.—Commencing at 54° 40', north latitude, ascending Portland channel to the mountains, following their summits to the 141° west longitude; thence north, on this line, to the Arctic ocean, forming the eastern boundary. Starting from the Arctic ocean west, the line descends Behring's strait, between the two islands of Krusenstern and Ratmanoff, to the parallel of 65° 30', and proceeds due north without limitation into the same Arctic ocean. Beginning again at the same initial point, on the parallel of 65° 30', thence in a course southwest through Behring's strait, between the island of St. Lawrence and Cape Chonkotski, to the 172° degree west longitude; and thence southwesterly through Behring's sea, between the island of Attou and Copper, to the meridian of 193° west longitude; leaving the prolonged group of the Aleutian islands in the possessions now transferred to the United States, and making the western boundary of our country the dividing line between Asia and America.

The above table shows but a small portion of the present domain to have been represented in the first Congress of the United States.

Tabular Statement showing the number of acres of public lands surveyed in the following land States and Territories up to June 30, 1866, during the last fiscal year, and the total of the public lands surveyed up to June 30, 1867; also the total area of the public domain remaining unsurveyed within the same.

Land States and Territories.	Area of the land States and Territories.		Number of acre of public lands surveyed up to June 30, 1866.	Number of acres of public lands surveyed during the fiscal year ending June 30, 1866, but not included in last year's report.	Number of acres of public lands surveyed within the fiscal year ending June 30, 1867.	Total of the public lands surveyed up to June 30, 1867.	Total area of public lands, including Russian territory, remaining unsurveyed, and of course unoffered and undisposed of; also private land claims surveyed and not reported up to June 30, 1867.
	In acres.	In square miles.					
Wisconsin	34,511,360	53,924	34,511,360			34,511,360	
Iowa	35,228,800	55,045	35,228,800			35,228,800	
Minnesota	53,459,840	83,531	22,045,867		884,690	22,910,612	30,549,228
Kansas	52,013,520	81,318	16,171,776	50	4,292,773	20,510,443	31,523,077
Nebraska	48,636,800	75,995	13,561,132	45,897	1,959,117	15,520,249	33,116,551
California	120,947,840	188,981	27,680,683	130,761	839,881	28,711,327	92,236,513
Nevada	71,737,741	112,090	728,119	19,531	16,319	763,969	70,973,772
Oregon	60,975,360	95,274	5,730,186		414,450	6,144,636	54,830,724
Washington Territory	44,796,160	69,994	3,530,645	54,843	294,550	3,880,038	40,916,122
Colorado Territory	66,880,000	104,500	1,622,251	91 821	1,130,775	2,844,887	64,035,143
Utah Territory	56,355,625	88,056	2,425,239		92,673	2,517,912	53,837,723

CHANGE OF NATIONAL EMPIRE. 67

Arizona Territory	72,906,304					72,906,304
New Mexico Territory	77,568,640	113,916				75,236,085
Dakota Territory	153,982,080	121,201				151,318,420
Idaho Territory	58,199,480	240,597				58,196,480
Montana Territory	92,016,640	50,932				92,016,640
Missouri	41,824,000	143,776	2,293,142	39,413	2,332,555	
Alabama	32,462,080	65,350	1,839,989	803,671	2,053,660	
Mississippi	30,179,840	50,722	41,824,000		41,824,000	
Louisiana	26,461,440	47,156	32,462,080		32,462,080	
Arkansas	33,406,720	41,346	30,179,840		30,179,840	3,000,000
Florida	37,931,520	52,198	23,461,440		23,461,440	
Ohio	25,576,960	59,268	33,406,720		33,406,720	
Indiana	21,637,700	39,964	25,631,520		26,631,520	11,300,000
Michigan	35,128,640	33,809	25,576,960		25,576,960	
Illinois	35,462,400	56,451	21,637,700		21,637,700	
Indian Territory	44,154,240	55,410	36,128,640		36,128,640	
		68,991	35,462,400		35,462,400	44,154,240
American purchase from Russia	369,529,600					369,529,600
Total	1,834,998,400	577,390	2,867,185	474,160,551	342,913	10,808,314	485,311,778	1,349,686,622

However considerable the figures may be, the preceding table shows that one-half our vast domain is not yet half surveyed, much less clothed with population, industry, wealth, and political power. These must come as the result of our national growth.

Since 1860 no part of our country has been growing so fast as that territorial portion which comprehends the mineral region; and, with the additional aid of the railway, the trade beyond the Mississippi has more than doubled in ten years, and in every direction Westward goes the American citizen in search of fame and fortune, and involuntarily adding to the greatness of the nation.

Everywhere on American soil or American water is seen the vitalizing principle of progress, urgent alike in every field of industry and every place of abode. Every new lesson of our country is of the great West, and every eulogy of the American statesman paints in glowing colors the Westward march of empire.

I have already stated that the internal commerce of the United States was greater than the external commerce. Not willing to enter thoroughly into the discussion of what may justly be termed the philosophy of the trade of the country and the influences that localize it, and thereby create commercial centers, I hereby submit certain papers, by J. W. Scott, Esq., formerly of Toledo, Ohio, which appeared in *Hunt's Merchants' Magazine* in 1843, 1848, and 1857. Mr. Scott was the editor of the *Toledo Blade* at the time of writing the articles, and, whatever else may be said of him, his articles show wonderful ability in the discussion of the commercial and material growth of the country. And, although in many respects he was mistaken, and the country has grown beyond his calculations, still his articles are worth reading, and are superior to those produced in any of the industrial and commercial magazines of the present time. He first speaks:

INTERNAL TRADE OF THE UNITED STATES.

Number I.—1843.

Almost up to the present time the whole weight of population in the United States has lain along the Atlantic shore, on and near its tide waters, and a great proportion of their wealth was connected with foreign commerce, carried on through their seaports. These being at once the centers of domestic and foreign trade, grew rapidly, and constituted all the large towns of the country. The inference was thence drawn, that as our towns of greatest size were connected with foreign commerce, this constituted the chief if not the only source of wealth, and that large cities could grow up nowhere but on the shores of the salt sea. Such had been the experience of our people, and the opinion founded on it has been pertinaciously adhered to, notwithstanding the situation of the country in regard to trade and commerce has essentially altered. It seems not, until lately, to have entered the minds even of well-informed statesmen that the internal trade of this country has become far more extensive, important, and profitable, than its foreign commerce. In what ratio the former exceeds the latter, it is impossible to state with exactness. We may, however, approximate the truth near enough to illustrate our subject.

The annual production of Massachusetts has been ascertained to be of the value of $100,000,000. If the industry of the whole nation were equally productive, its yearly value would be about $2,300,000,000; but, as we know that capital is not so abundantly united with labor in other States, it would be an over-estimate to make that State a basis of a calculation for the whole country. $1,500,000,000 is probably near the actual amount of our yearly earnings. Of this, there may be $500,000,000 consumed and used where it is earned, without being exchanged. The balance, being $1,000,000,000, constitutes the subjects of exchange, and the articles that make up the domestic trade and foreign commerce of the United States. The value of those which enter into our foreign commerce is, on an average, about $100,000,000. The average domestic exports of the years 1841 and 1842 is $99,470,900. There will then remain $900,000,000, or nine-tenths, for internal trade. Supposing, then, some of our towns to be adapted only to foreign commerce, and others as exclusively fitted for domestic trade; the latter, in our country, would have nine times as much business as the former, and should, in consequence, be nine times as large. Although we have no great towns that do not, in some degree, participate in both foreign and domestic trade, yet we have those whose situations particularly adapt them to one or the other; and we wish it constantly borne in mind that an adaptation to internal trade, other things being equal, is worth nine times as much to a town as an adaptation in an equal degree to foreign commerce. It may be said, and with truth, that our great seaports have manifest advantages for domestic as well as foreign commerce. Since the peace of Europe left every nation free to use its own navigation, the trade of our Atlantic coast has probably been five times greater than that carried on with foreign nations; as the coasting tonnage has exceeded the foreign, and the number of voyages of the former can scarcely be less than five to one of the latter.

Now, what is the extent and quality of that coast, compared with the navigable river and the coasts of the North American valley?* From the mouth of the St. Croix to Sandy Hook, the soil, though hard and comparatively barren, is so well cultivated as to furnish no inconsiderable amount of products for internal trade. In extent, including bays, inlets, and both shores and navigable rivers, and excluding the sand beach known as Cape Cod, this coast may be estimated at 900 miles. From Sandy Hook to Norfolk, including both shores of Delaware and Chesapeake bays, and their navigable inlets, and excluding the barren shore to Cape May, the coast may be computed at 900 miles more. And from Norfolk to the Sabine there is a barren coast of upwards of 2,000 miles, bordered most of the way by a sandy desert extending inland on an average of 80 or 90 miles. Over this desert must be transported most of the produce and merchandise, the transit and exchange of which constitute the trade of this part of the coast. This barrier of nature must lessen its trade at least one-half. It will be a liberal allowance to say that 4,000 miles of accessible coast are afforded to our vessels by the Atlantic Ocean and Gulf of Mexico. Of this, only about 2,500 miles, from Passamaquoddy to St. Mary's, can be said to have contributed much, until recently, to the building of our Atlantic cities. To the trade of this coast, then, are we to attribute five-sixths of the growth and business, previous to the opening of the Erie canal, of Portland, Salem, Boston, Providence, New York, Albany, Troy, Philadelphia, Baltimore, Washington, Richmond, Norfolk, Charleston, Savannah, and several other towns of less importance. Perhaps it will be said that foreign trade is more profitable, in proportion to its amount, than domestic. But is this likely? Will not the New York merchant be as apt to make a profitable bargain with a Carolinian as with an Englishman of Lancashire? Or, is it an advantage to trade to have the wide obstacle of the Atlantic

---

* This valley includes the basins of the St. Lawrence, Mississippi and Mobile rivers.

in its way? Do distance and difficulty, and risk and danger, tend to promote commercial intercourse and profitable trade? If so, the Alleghanies are a singular blessing to the commercial men living on their western slope. Some think that it is the foreign commerce that brings all the wealth to the country, and sets in motion most of the domestic trade. At best, however, we can only receive by it imported values, in exchange for values exported, and those values must be first created at home.

With the exception of tobacco, our exports to foreign nations are mostly prime necessaries of life, such as minister in the highest degree to the comforts of the people who use them. Such are breadstuffs, provisions, and cotton-wool, a material from which a great part of the clothing of the world is fabricated. And what do we receive in exchange so calculated to enrich us as a nation? Among other articles imported in 1840 (we have not before us a later return from the Secretary of the Treasury) we received tea and coffee, to the value of (we give round numbers) $14,000,000; silks, and silk and worsted stuffs, near $10,250,000; wines and spirits, $3,600,000; lace, $500,000; tobacco, manufactured, $870,000; in all, near $30,000,000 out of an import of $107,000,000. The dealing in these articles may have a tendency to enrich, but surely neither those that consume, nor those whose labor buys, the articles above specified, are enriched. Indeed, if the $300,000,000 of food and materials for clothing, which are sent abroad to pay for such poisons and luxuries, are not wholly lost by being so exchanged, it will be admitted that we are not greatly enriched by the exchange. Let us not be understood as desirous of undervaluing foreign trade. We hope and believe that its greatest blessings and triumphs are yet to come. Many of the articles which it brings to us add much to our substantial comfort, such as woolen and cotton goods, sugar and molasses; and others, such as iron and steel, with most of their manufactures, give much aid to our advancing arts. But if these articles were the products of domestic industry — if they were produced in the factories of Lowell and Dayton, on the plantations of Louisiana, and in the furnaces, forges, and workshops of Pennsylvania — why would not the dealing in them have the same tendency to enrich as now that they are brought from distant countries?

A disposition to attribute the rapid increase of wealth in commercial nations mainly to foreign commerce, is not peculiar to our nation or our time; for we find it combated as a popular error by distinguished writers on political economy. Mr. Hume, in his Essay on Commerce, maintains that the only way in which foreign commerce tends to enrich a country is by its presenting tempting articles of luxury, and thereby stimulating the industry of those in whom a desire to purchase is thus excited; the augmented industry of the nation being the only gain.

Dr. Chalmers says: "Foreign trade is not the creator of any economic interest; it is but the officiating minister of our enjoyments. Should we consent to forego those enjoyments, then, at the bidding of our will, the whole strength at present embarked in the service of procuring them would be transferred to other services, to the extension of the home trade; to the enlargement of our national establishments; to the service of defense, or conquest, or scientific research, or Christian philanthropy." Speaking of the foolish purpose in Bonaparte to cripple Britain by destroying her foreign trade, and its utter failure, he says: "The truth is that the extinction of foreign trade in one quarter was almost immediately followed up either by the extension of it in another quarter, or by the extension of the home trade. Even had every outlet abroad been obstructed, then, instead of a transference from one foreign market to another, there would just be a universal reflux towards a home market that would be extended in precise proportion with every successive abridgment which took place in our external commerce." If these principles are true, and we believe they are in accordance with those of every eminent writer on political economy, and if they are important in their application to the British isles — small in territory — with extensive districts of barren land — surrounded by navigable waters — rich in good harbors, and presenting numerous natural obstacles to constructions for the promotion of internal commerce; and, moreover, placed at the door of the richest nations of the world — with how much greater force do they apply to our country, having a territory twenty times as large, unrivaled natural means of intercommunication, with few obstacles to their indefinite multiplication by the hand of man; a fertility of soil not equaled by the whole world; growing within its boundaries nearly all the productions of all the climes of the earth, and situated 3,000 miles from her nearest commercial neighbor.

Will it be said that, admitting the chief agency in building up great cities to belong to internal industry and trade, it remains to be proved that New York and the other great Atlantic cities will feel less of the beneficial effects of this agency than Cincinnati and her Western sisters? It does not appear to us difficult to sustain by facts and reasoning the superior claims in this respect of our Western towns. It should be borne in mind that the North American Valley embraces the climate, soils, and minerals, usually found distributed among many nations. From the northern shores of the upper lakes, and the highest navigable points of the Mississippi and Missouri rivers, to the Gulf of

Mexico, nearly all the agricultural articles which contribute to the enjoyment of civilized man are now, or may be, produced in profusion. The North will send to the South grain, flour, provisions, including the delicate fish of the lakes, and the fruits of a temperate clime, in exchange for the sugar, rice, cotton, tobacco, and the fruits of the warm South. These are but a few of the articles, the produce of the soil, which will be the subjects of commerce in this valley. Of mineral productions, which, at no distant day, will tend to swell the tide of internal commerce, it will suffice to mention coal, iron, salt, lead, lime, and marble. Will Boston, or New York, or Baltimore, or New Orleans, be the point selected for the interchange of these products? Or, shall we choose some convenient central points on river and lake for the theaters of these exchanges? Some persons may be found, perhaps, who will claim this for New Orleans; but the experience of the past, more than the reason of the thing, will not bear them out. Cincinnati has now more white inhabitants than that outport, although her first street was laid out, and her first log house raised, long after New Orleans had been known as an important place of trade, and had already become a considerable city.

It is imagined by some that the destiny of this valley has fixed it down to the almost exclusive pursuit of agriculture, ignorant that, as a general rule in all ages of the world, and in all countries, the mouths go to the food, and not the food to the mouths. Dr. Chalmers says: "The bulkiness of food forms one of those forces in the economic machine which tend to equalize the population of every land with the products of its own agriculture. It does not restrain disproportion and excess in all cases; but in every large State it will be found that wherever an excess obtains, it forms but a very small fraction of the whole population. Each trade must have an agricultural basis to rest upon; for in every process of industry, the first and greatest necessity is, that the workmen shall be fed." Again: "Generally speaking, the *excrescent* (the population over and above that which the country can feed) bears a very minute proportion to the natural population of the country; and almost nowhere does the commerce of a nation overleap, but by a very little way, the basis of its own agriculture." The Atlantic States, and particularly those of New England, claim that they are to become the seats of the manufactures with which the West is to be supplied; that mechanics, and artisans, and manufacturers, are not to select for their place of business the region in which the means of living are most abundant and their manufactured articles in greatest demand, but the section which is most deficient in those means, and to which their food and fuel must, during their lives, be transported hundreds of miles, and the products of their labor be sent back the same long road for a market.

But this claim is neither sanctioned by reason, authority, nor experience. The mere statement exhibits it as unreasonable. Dr. Chalmers maintains that the "excrescent" population could not, in Britain even, with a free trade in breadstuffs, exceed one-tenth of all the inhabitants; and Britain, be it remembered, is nearer the granaries of the Baltic than is New England to the food-exporting portions of our valley, and she has, also, greatly the advantage in the diminished expense of transportation. But the Eastern manufacturing States have already nearly, if not quite, attained to the maximum ratio of excrescent population, and cannot, therefore, greatly augment their manufactures without a correspondent increase in agricultural production.

Most countries, distinguished for manufactures, have laid the foundation in a highly improved agriculture. England, the north of France, and Belgium, have a more productive husbandry than any other region of the same extent. In these same countries are also to be found the most efficient and extensive manufacturing establishments of the whole world; and it is not to be doubted that abundance of food was one of the chief causes of setting them in motion. How is it that a like cause operating here will not produce a like effect? Have we not, in addition to our prolific agriculture, as many and as great natural aids for manufacturing as any other country? Are we deficient in water-power? Look at Niagara river, where all the accumulated waters of the upper St. Lawrence basin fall *three hundred and thirty-five feet* in the distance of a few miles. Ohio, or Kentucky, or Western Virginia, or Michigan, can alone furnish durable water-power, far more than sufficint to operate every machine in New England. The former State has now for sale on her canals more water-power than would be needed for the moving of all the factories of New England and New York. Indeed, no idea of our Eastern friends is more preposterous than the one so hugged by them, that they of all the people of the Union are peculiarly favored with available water-power. We remember reading in the *North American Review*, many years ago, in an article devoted to the water-power, and its appropriation in the neighborhood of Baltimore, that southwardly from that city the Atlantic States were destitute of water-power; when every well-informed man should know that there is not one of those States in which its largest river would not furnish more than power sufficient to manufacture every pound of cotton raised within its boundaries. The streams of New England are short and noisy, not an unfit emblem of her manufacturing pretensions and destiny.

But if our water-power should be unequal to our manufacturing exigencies, our beds of coal will not fail us. One of these coal formations, having its center not very far from Marietta, is estimated by Mr. Mather, geologist, to be of the extent of 50,000 square miles. He says that in several of the counties of Ohio the beds of workable coal are from 20 to 30 feet thick. Another coal formation embraces the Wabash Valley of Indiana, and the Green river country of Kentucky. We know also of its existence in abundance at Ottawa and Alton, in the State of Illinois, and suppose they are in the same coal basin. Another coal basin has been discovered in Michigan, and a fifth on the Arkansas river. In some of these coal regions, and probably in all, beds of iron ore and other valuable minerals for manufacture are abundant.

Will laborers be wanting? Where food is abundant and cheap, there cannot long be a deficiency of laborers. What brought our ancestors (with the exception of a few who fled from persecution) from the other side of the Atlantic, but the greater abundance of the means of subsistence on this side? What other cause has so strongly operated in bringing to our valley the 10,000,000 or 11,000,000 who now inhabit it? The cause continuing, will the effect cease? While land of unsurpassed fertility remains to be purchased, at a low rate, and the increase of agriculture in the West keeps down the relative price of food; and while the population of the old countries of Europe and the old States of our confederacy is so augmenting as to straiten more and more the means of living at home, and at the same time the means of removing from one to the other are every year rendering it cheaper, easier, and more speedy; and while, moreover, the new States, in addition to the inducement of cheaper food, now offer a country with facilities of intercourse among themselves greatly improved, and with institutions, civil, political, and religious, already established and flourishing — are farmers, and mechanics, and manufacturers — the young, the active, and the enterprising — no longer to be seen pouring into this exuberant valley and marking it with the impress of their victorious industry, as in times past?

If our readers are satisfied that domestic or internal trade must have the chief agency in building up our great American cities, and that the internal trade of the great Western valley will be mainly concentrated in the cities situated within its bosom, it becomes an interesting subject of inquiry how our leading interior city will, at some distant period, say 100 years, compare with New York, the Atlantic emporium. For the purpose of illustration, let us take Cincinnati as the chief interior city. Whether it will actually become such, we design to discuss in a separate paper.

One hundred years from this time, if our ratio of increase for the last 50 years is kept up, our Republic will number, in round numbers, 325,000,000—say 300,000 000. Of this number, if we allow for the Atlantic slope five times its present population, or 40,000,000, and to the Oregon country 10,000,000, there will remain for our great valley 250,00 ,000. If to these we add the 20,000,000 by that time possessed by Canada, we have, for our North American valley, 270,000,000. The point, then, will be reduced to the plain and easily solved question, whether 270,000,000 of inhabitants will build up and sustain greater cities than 40,000,000. As our valley is in shape more compact than the Atlantic slope, it is more favorable to a decided concentration of trade to one point. Whether that point is most likely to be Cincinnati, or some rival on the lake border, we propose hereafter to consider.

Let us now see what facilities for internal commerce nature has bestowed on the West It will not be denied that, for internal trade, the country bordering the Ohio, Mississippi, and other rivers admitting steam navigation, are at least as well situated as if laved by the waters of an ocean. Cincinnati being at present the leading city of our valley, we propose to connect it particularly with our argument, not doubting that other and many great towns will grow up on the Western waters. From Pittsburgh to Cincinnati, both shores of the Ohio amount to more than 900 miles. From Cincinnati to New Orleans, there is a river coast of 3,000 miles The upper Mississippi has 1,600 miles of fertile shore. The shores of that part of the Missouri which has been navigated by steamboats amount to near 4,000 miles. The Arkansas, Red, Illinois, Wabash, Tennessee, Cumberland, St. Francis, Wh te, Ouachita, have an extent of shore, accessible to steamers, of not less than 8,000 miles.

Here, then, are fertile shores, to the extent of near 20,000 miles, which can be visited by steam-vessels a considerable part of the year. Taking these streams together, they probably afford facilities for trade nearly equal in value to the same number of miles of common canals. Who, then, can doubt that in the midst of such facilities for trade large cities must grow up, and with a rapidity having no example on the Atlantic coast. The growth of Cincinnati, Pittsburgh, Louisville, and St. Louis, since 1825, gives us abundant assurances on this point.

But our interior cities will not depend for their development altogether on internal trade. They will partake, in some degree, with their Atlantic sisters, of the foreign commerce, also; and if, as some seem to suppose, the profits of commerce increase with the distance at which it is carried on, and the difficulties which nature has thrown in its way, the Western towns will have the same advantage over their Eastern rivals in foreign commerce, which some claim for the latter over the former in our domestic trade. Cincinnati and her lake rivals may use the outports of New Orleans and New York, as Paris and Vienna use those of Havre and Trieste; and it will surely one day come to pass that steamers from Europe will enter our great lakes, and be seen booming up the Mississippi.

To add strength and conclusiveness to the above facts and deductions, do our readers ask for examples? They are at hand. The first city of which we have any record is Nineveh, situated on the Tigris, not less than 700 miles from its mouth. Babylon, built not long after, was also situated far in the interior, on the river Euphrates. Most of the great cities of antiquity, some of which were of immense extent, were situated in the interior, and chiefly in the vallies of large rivers, meandering through rich alluvial territories. Such were Thebes, Memphis, Ptolemais. Of the cities now known as leading centers of commerce, a large majority have been built almost exclusively by domestic trade. What country has so many great cities as China, a country, until lately, nearly destitute of foreign commerce?

To bring the comparison home to our readers, we have put down, side by side, the outports and interior towns of the world having each a population of 50,000 and upwards. It should, however, be kept in mind that many of the great seaports have been built, and are now sustained, mainly by the trade of the nations respectively in which they are situated. Even London, the greatest mart in the world, is believed to derive much the greatest part of the support of its vast population from its trade with the United Kingdom.

OUTPORTS.	Population.	INTERIOR CITIES.	Population.	INTERIOR CITIES.	Population.
London	2,000,000	Pekin	1,300,000	Florence	80,000
Jeddo (?)	1,300,000	Paris	1,000,000	Gallipolis	80,000
Calcutta	650,000	Benares	600,000	Bucharest	80,000
Cons'tinople	600,000	Hang-tcheou	600,000	Munich	80,000
St. Pet'sburgh	500,000	Su-tcheon	600,000	Granada	80,000
Canton (?)	500,000	Macao	500,000	Ghent	80,000
Madras	450,000	Nankin	500,000	Lassa	80,000
Naples	350,000	Bing-tchin	500,000	Cologne	75,000
Dublin	330,000	Woo-tchang	400,000	Morocco	75,000
New York	320,000	Vienna	370,000	Ferruckabad	70,000
Lisbon	250,000	Cairo	350,000	Peshawen	70,000
Glasgow	250,000	Patna	320,000	Quito	70,000
Liverpool	250,000	Nan-tchang	300,000	Barreilly	70,000
Philadelphia	250,000	Khai-fung	300,000	Guadalaxara	70,000
Rio Janeiro	200,000	Fu-tchu	300,000	Koenigsburg	70,000
Amsterdam	200,000	Lucknow	300,000	Turgan	70,000
Bombay	200,000	Moscow	300,000	Salonica	70,000
Palermo	170,000	Berlin	300,000	Bologna	70,000
Surat	160,000	Manchester	250,000	Bornaserai	70,000

CHANGE OF NATIONAL EMPIRE.

OUTPORTS.	Population.	INTERIOR CITIES.	Population.	INTERIOR CITIES.	Population.
Manilla	140,000	Birmingham	230,000	Dresden	70,000
Hamburg	130,000	Lyons	200,000	Lille	70,000
Bristol	120,000	Madrid	200,000	Norwich	70,000
Havana	160,000	Delhi	200,000	Perth	70,000
Marseilles	130,000	Aleppo	200,000	Santiago	60,000
Barcelona	120,000	Mirzapore	200,000	Wilna	60,000
Copenhagen	120,000	Hyderbad	200,000	Cabul	60,000
Smyrna	120,000	Dacca	200,000	Khokhan	60,000
St. Salvador	120,000	Ispahan	200,000	Samarcand	60,000
Cork	120,000	Yo-tchu	200,000	Resht	60,000
Brussels	120,000	Suen-tchu	200,000	Casween	60,000
Bordeaux	100,000	Huen-tchu	200,000	Diarbekir	60,000
Venice	100,000	Mexico	200,000	Karahissar	60,000
Baltimore	100,000	Leeds	180,000	Mosul	60,000
New Orleans	100,000	Lyons	180,000	Bassora	60,000
Boston	100,000	Moorshedabad	160,000	Mecca	60,000
Tunis	100,000	Milan	160,000	Mequirez	60,000
Nantes	100,000	Damascus	150,000	Bungalore	60,000
Hue	100,000	Cashmere	150,000	Bardwan	60,000
Bankok	90,000	Rome	150,000	Anrangabad	60,000
Seville	90,000	Edinburgh	150,000	Nottingham	60,000
Gallipoli	80,000	Teheran	130,000	Oldham	60,000
Genoa	80,000	Turin	120,000	Cordova	57,000
Stockholm	80,000	Prague	120,000	Verona	56,000
Newcastle	80,000	Warsaw	120,000	Padua	55,000
Massalipatan	75,000	Sheffield	120,000	Frankfort	54,000
Pernambuco	75,000	Bagdad	100,000	Liege	54,000
Lima	75,000	Brussa	100,000	Lemberg	52,000
Greenwich	75,000	Tocat	100,000	Stoke	52,000
Aberdeen	70,000	Erzeroum	100,000	Kazar	50,000
Antwerp	70,000	Poonah	100,000	Salford	50,000
Limerick	70,000	Nagpore	100,000	Strasburg	50,000
Valentia	65,000	Ahmedabad	100,000	Amiens	50,000
Rotterdam	65,000	Lahore	100,000	Kutaiah	50,000
Leghorn	65,000	Baroda	100,000	Trebizond	50,000
Dantzic	65,000	Orogein	100,000	Orfa	50,000
Batavia	60,000	Candahar	100,000	Tariga	50,000
Cadiz	55,000	Balfrush	100,000	Cuzco	50,000
Hull	55,000	Herat	100,000	Puebla	50,000
Belfast	55,000	Saigon	100,000	Metz	50,000
Portsmouth	55,000	Breslau	100,000	Hague	50,000
Trieste	55,000	Adrianople	100,000	Bath	50,000
Malaga	52,000	Kesho	100,000	Constantina	50,000
N. Guatimala	50,000	Rouen	100,000	Cairwan	50,000
Muscat	50,000	Toulouse	90,000	Gondar	50,000
Algiers	50,000	Indore	90,000	Ava	50,000
Columbo	50,000	Wolverh'pton	90,000	Rampore	50,000
Odessa	50,000	Paisley	90,000	Mysore	50,000
		Jackatoo	80,000	Bardwar	50,000
		Tauris	80,000	Boli	50,000
		Bucharia	80,000	Hamah	50,000
		Gwallior	80,000	Cincinnati	50,000

If it be said that the discoveries of the polarity of the magnetic needle, the continent of America, and a water passage to India, around the Cape of Good Hope, have changed the character of foreign commerce, and greatly augmented the advantages of the cities engaged in it, it may be replied that the introduction of steam in coast and river navigation, and of canals, and railroads, and McAdam roads, all tending to bring into rapid and cheap communication the distant parts of the most extended continent, is a still more potent cause in favor of internal trade and interior towns. The introduction, as instruments of commerce, of steamboats, canals, rail, and McAdam roads, being of recent date, they have not had time to produce the great results that must inevitably flow from them. The last 20 years have been devoted mainly to the construction of these labor-saving instruments of commerce; during which time more has been done to facilitate internal trade than had been effected for the thousands of years since the creation of man. These machines are but just being brought into use; and he is a bold man who, casting his eye 100 years into the future, shall undertake to tell the present generation what will be their effect on our North American valley when their energies shall be brought to bear over all its broad surface.

Let it not be forgotten that, while many other countries have territories bordering the ocean, greatly superior to our Atlantic slope, no one government has an interior at all worthy a comparison with ours. It will be observed that, in speaking of the natural facilities for trade in the North American valley, we have left out of view the 4,000 or 5,000 miles of rich and accessible coasts of our great lakes and their connecting straits. The trade of those inland seas, and its connection with that of the Mississippi Valley, are subjects too important to be treated incidentally in an article of so general a nature as this. They well merit a separate notice at our hands.

Number II.—1843.

Providence has evidently designed the temperate regions of the interior of North America for the residence of a dense population of highly civilized men. Throughout its southern and middle regions, which are elevated but a few hundred feet above the level of the Gulf of Mexico, the deflected trade wind bears from that sea the vapors which, falling in showers, give fertility to the soil, and swell to navigable size their numerous and almost interminable rivers. Towards the North he has spread out, and connected by navigable straits, great seas of pure water, to equalize and soften the temperature of that comparatively high latitude, and to aid in irrigating the surrounding countries. And he has so placed these seas as to give them the utmost availability for purposes of trade; for, while they reach to the highest latitude to which profitable cultivation can be carried, they stretch away South almost to the very heart of the great valley. Towards the East they approach the Atlantic, and extend Westward towards the Pacific, more than a third of the distance across the continent. To give the lake and river countries easy access to each other, he has placed them nearly on the same level, and strongly pointed out, and, indeed, in some places, almost finished, the great channels of intercourse between them. To invite and facilitate migration from Europe and the old States, he has provided the St. Lawrence and Mississippi rivers, and cut a passage through the Appalachian chain, where flow the turbulent Mohawk and the majestic Hudson. His munificence ends not here. He has diversified its surface with hills, vales, and plains, and clothed them alternately with fine groves of timber and beautiful meadows of grass and flowers. Beneath the soil, the minerals of nearly every geological era, and of every kind which has been made tributary to man's comfort and civilization, are properly distributed. On the north, the waters of the great lakes begin their expansion in a region of primitive formation. Descending thence by the river St. Mary's into, and expanding over, a portion of that great transition limestone bed which forms the basis of the richest soil of the country, and after entering, by their southernmost reach, the coal measures of northern Ohio, they are precipitated over the eastern margin of this great limestone basin at Niagara. A few miles distant they again spread out, 330 feet below, in a region of salt-bearing sandstone and shales, and finally pass off to the ocean through a primitive country. Thus a great variety of minerals, useful to man, are placed where transportation and exchange are easy and cheap. Nor, in this connection, should be overlooked, among the multiplied evidences of Providential bounties to this favored region, the immense power to move machinery laid up for us at the outlet of Lake Erie. Here is a head of 330 feet, with an inexhaustible supply of pure water, easily and cheaply brought under control, in a healthy and pleasant country, and at the door of the great West. Nor should we omit to mention the harbors for the shipping, which abound in the primitive shores to the North, and which are also found at the mouths of all large streams of the transition and secondary region below.

Such is the broad patrimony which we are invited to enter upon and improve. Our people have begun to enter into possession. Along the line of the 5,000 or 6,000 miles of habitable shore which is offered to the mariner of these lakes, he may now and then see a cluster of houses, a nascent city; and anon he may espy small indentations of their forest borders, where farmers have begun to hew their way to independence. The southern shore of Lake Erie, and both shores of Ontario, are so far advanced in settlement that it is easy to anticipate the speedy triumph of the art and industry of man. Already, in many places, he has achieved his victory; for his farms and villages have nearly driven his forest enemy from his sight. Here he has already built himself spacious barns and comfortable dwellings. He has also made roads on which to carry the produce of his industry to market. More than this: he has built towns, canals, and railroads, constructed and improved numerous harbors, and created a commercial marine that, three centuries ago, would have been a source of pride if possessed by the greatest maritime power in Europe.

In anticipation of the early settlement of the fine country bordering these waters, and its capacity to furnish the basis of a large commerce, the Erie Canal was projected and opened. But its banks had hardly become solid, its business been got into train and reduced to system, before the discovery was made that its capacity would little more than suffice for the business of the country through which it runs, and, of course, that it would soon be inadequate to the passage of the trade then just springing up, with indications of a vigorous growth, on the upper lakes. Wild as were thought the visions of Morris and Clinton by the strictly practical men of their day, it turns out that what were considered *visions* were but practical deductions, falling short of the truth instead of exceeding it. Ten years after the *chimerical* grand canal was completed, men, having the reputation of being eminently practical, thought they saw the necessity of making it about three times as large, and forthwith entered upon such enlargement. Practical men in other States have believed, perhaps prematurely, that such portion of the lake trade as they could divert from this New York route would pay them for the outlay of so many millions as will be necessary to construct two more canals, and the same number of railroads, from the Alantic to the lake waters. Not only are cities and States entering upon a competition for this trade, but there are indications that a few years will witness an active emulation between the United States and Great Britain, in endeavors, on the one hand, to retain, and, on the other, to acquire it. On all sides it is admitted that the city of the Atlantic coast which receives the bulk of our Easten business will be the leading city of that border; and if it is not now admitted, it will soon be, that the emporium of the Mississippi Valley which commands the best channel of intercourse with the lakes must be, and remain, the queen city of the valley. But what is it that makes this lake country of such commanding importance? In the first place, it is of great extent. Its navigable shores, including bays and straits, measure more than 5,000 miles. Not only do these command a large country lying back, in many places, much beyond the head waters of the streams which flow into them, but, by means of valleys, canals, and other artificial aids, no inconsiderable portion of the Mississippi Valley is made tributary to their commerce. This is owing to their affording the cheapest and best route to New York and Canada. Even with the small canal between Buffalo and Albany, levying tolls high enough to have already paid for its construction, we find a strong inclination to that route, not only for the foreign and Eastern manufactures that are purchased in the great Atlantic emporium, and brought into the lake and Mississippi valleys, but for the farming produce of sections of country that formerly floated it down to New Orleans. This is strongly exemplified on the Ohio Canal, the lake end of which receives of the agricultural productions transported on it more than twelve times as much in value as the Ohio river termination. We have examined the receipts by canal, at Cleveland and Portsmouth, for the six past years — the only years for which the board of public works have given full returns — and the result shows the above proportion. For those six years,

Cleveland received of wheat..................................8,325,022 bushels.
Portsmouth   "   "   4,193   "
Cleveland   "   "   flour..........................2,190,542 barrels.
Portsmouth   "   "   149,645   "

When the Erie Canal shall be made three times its original size, through its whole length, to Buffalo, or from Albany to Syracuse, with an equivalent enlargement of the Oswego Canal, the cost of transportation on it will be materially diminished, so as to draw trade to the lakes from a still more extended portion of the great valley. This tendency will be increased by the facilities which the Canadian improvements will give the lake ports, to make shipments direct to foreign ports; and it will, in like manner, be greatly strengthened by the completion of the Wabash and Erie Canal, which comes first into operation the present season; and by the Miami Canal, which will connect Cincinnati with the lake, by a direct communication of only 235 miles in length, and which will be in operation in the summer of 1844. Until the cities and towns of the central valley become numerous and large enough to consume most of its agricultural surplus, the main exertions of her people will be properly directed to the construction and improvement of channels for its transport, by way of the lakes, to Quebec, New York, and Boston.

The country lying north and northwest of the lakes, to an almost infinite extent, must carry on its main exchanges through these waters. This, though new and little improved, will, at no very distant day, become populous and powerful. Before the late troubles, the migration to Upper Canada from the United Kingdom was unexampled in the history of colonization, being, some seasons, upwards of 50,000 annually. Quiet being again restored, the current in that direction is becoming stronger than ever.

The soil of the countries bordering the lakes is, in general, of the most fertile character; and the climate, for health and pleasantness, equal to that of any part of the continent, except, perhaps, the table lands of Mexico. They join, and are in the same latitude, with those Atlantic States having the densest population and the greatest wealth; and the expenditure of time and money to change a residence from these to the lake borders is now small, and is every year becoming less. The main current of surplus population has for several years flowed from those States into the lake region; and that current will grow wider, and deeper, and stronger, in proportion to the removal of obstacles impeding its progress.

Now let us see what means are in course of preparation for making easy and cheap the intercourse between the lakes and the Atlantic States. First in importance is the enlarged Erie Canal. This work is now in progress, and it will probably be finished, as far as its connection with the Oswego Canal at Syracuse, in two years. By that time, it is hoped, the Oswego branch will also be enlarged to the same size. Its dimensions are 70 feet in width, 7 feet in depth, with double locks throughout, large enough to pass vessels of 150 tons.

Next in importance, when finished, will be the Chesapeake and Ohio Canal, with its continuation from Pittsburgh to Cleveland. This will be a continuous line of canal, about 520 miles in length, connecting tide water at Baltimore, and Georgetown with Lake Erie, at Cleveland. Its dimensions vary from 40 feet wide and 4 feet deep to 60 feet wide by 6 feet deep; averaging, say 50 feet wide and 5 feet deep.

The Pennsylvania line of canal and railroad will unite with the foregoing at Pittsburgh, and from tide water at Philadelphia to Cleveland will be about 570 miles long. These are the rival canal routes in the States for the trade of the lakes. Let them stand together, that we may see how they compare:

	Length. Miles.	Size. Feet.	Lockage. Feet.	Tr'shipm. No.
1. Erie Canal, from Buffalo to Albany	363	70 by 7	688	None.
2. Chesapeake and Ohio, and Mahoning and Ohio Canal to Cleveland	520	50 by 5	4,500	3
3. Pennsylvania Works, and Mahoning and Ohio Canal, to Cleveland	570	40 by 4	5,000	3

It is a contrast rather than a comparison. If, however, the other routes were to afford equal facilities for business, that to New York would have a decided preference, because it leads to that established and controlling mart. But the Erie Canal is to have a formidable foreign rival. Canals are in process of construction around the rapids of St. Lawrence, of a size, and with locks, large enough to admit large steamboats; and the Welland Canal and locks are also being made capable of passing small steam vessels, and sailing vessels of 300 tons. These, when completed, will give entrance at once to foreign vessels of 1,000 tons burden to Lake Ontario, and of 300 tons to the ports of Lake Erie. These works are vigorously going forward to completion, the money necessary for that purpose being pledged under a guarantee of the home government. Many expect them to be finished in about two years; but we fear this expectation is over-sanguine. A comparison of the New York and Canada routes would stand thus:

From Lake Erie to New York, by canal and Hudson River—

Distance. Miles.	Size of Canal. Feet.	Size of Locks. Feet.	Length of Canal. Miles.	Lake and River. Miles.	L'kage. Feet.	No of Tr'shipm.
508	70 by 7	120 by 24	360	145	688	1

From entrance of Welland Canal on Lake Erie, to Montreal—

| 407 | 100 by 10 | 200 by 50 | 60¼ | 346 | 517 | None. |

The locks of the Welland Canal are being constructed 122 feet long in the chamber and 26 feet wide. It will be seen that we have set down the size of the Erie Canal as if enlarged all the way to Lake Erie; and the size of the Canadian locks, on the St. Lawrence, as if continued to the same lake. We have set down but one transhipment against the New York route by Buffalo; whereas, in regard to all freights coming from other ports of the upper lakes, there will, of course, be a reshipment at Buffalo, as well as at Troy or Albany. Let us see how the New York route, by Oswego, will compare with that of the St. Lawrence:

From exit of Welland Canal, in Lake Ontario, to New York—

Distance.	Size of Canal.	Size of Locks.	Length of Canal.	Lake and River.	Lockage.	Reshipments.
504 miles.	70 by 7 feet.	120 by 24 feet.	209 miles.	295 miles.	551 feet.	2

From exit of Welland Canal, in Lake Ontario, to Montreal—
379 miles. 110 by 10 ft. 200 by 50 feet. 32¼ miles. 347 miles. 188½ ft. None.

In a report of the Board of Directors of the Welland Canal, in 1835, it is stated that "merchandise from London would be conveyed to Cleveland for £2 10s. per ton," when the St. Lawrence should be rendered navigable to the lakes by the works now in process of construction. This would be 54 cents per 100 lbs., not above two-thirds its present cost from New York. If this statement be not greatly erroneous, European goods will be delivered at the ports of Lake Erie, on the completion of the Canadian canals, cheaper than at the port of New Orleans.

The railroads made, and in progress, to connect the ocean and the lakes, are: 1st, that from Buffalo to Albany, and thence by branches to Boston, New York, and all the large towns of New England and the State of New York; 2d, the Hudson and Erie, from Dunkirk to the Hudson; 3d, the Sunbury, from Erie to Philadelphia, and 4th, the Baltimore and Ohio, which, beginning at Baltimore and Washington, will, one day, terminate on Lake Erie, at Cleveland and Maumee; the former branch passing through Pittsburgh, the latter through Wheeling. Of these routes, that passing along near the line of the Erie Canal possesses nearly the same advantage over the others, as that canal has been shown to afford over her would-be rivals of Pennsylvania and Maryland. It avoids the ascent and descent of the Alleghany Mountains, and, passing along a level country, is much straighter, is made and kept in repair at much less expense, and, of course, will allow a greater speed to the locomotives that fly along its track.

Such are the great works made and making; and for whom? Surely not for the two or three millions that, within a few years past, have fixed their home in the lake countries. No! but for the anticipated tens of millions of intelligent and industrious freemen, who will, as a moderate forecast enables men to see, in no long course of years, spread over and clear and cultivate and beautify these pleasant and fertile shores. Whatever other error may arise from making the past a basis of calculation for the future, that of a too sanguine estimate could hardly be committed, in treating of any civilized country of the present day, much less of ours, the most rapidly progressive of the whole family of nations. To exhibit the growth of the principal upper lake towns, from 1830 to 1840, we here give their population at those periods:

	1830.	1840.		1830.	1840.
Buffalo	8,653	18,213	Detroit	2,222	9,102
Erie	1,329	3,412	Monroe	500	1,703
Cleveland *	1,076	7,648	Chicago	100	4,470
Sandusky City	400	1,433	Milwaukee	20	1,712
Lower Sandusky	351	1,117	Huron	75	1,488
Perrysburg	182	1,065			
Maumee City	200	1,290		2,917	18,475
Toledo	30	2,053		12,221	36,231
	12,221	36,231	Total	15,138	54,706

* Including Ohio City.

Showing an increase which, if the numerous villages that have commenced their existence since 1830 were added, would more than quadruple their numbers in ten years. The increase of business on the upper lakes has been in a greater ratio than even ten to one. Indeed, it has nearly all grown up since 1830. If the reader doubt this, let him examine and compare the account of the collector of canal tolls at Buffalo for that year with that for the past season, and add to the last the produce passing through the Welland Canal.

But it should not be forgotten that, while the relative amount of produce of the soil, in proportion to the population, is rapidly augmenting, our cities and towns are beginning to receive a large accession of mechanics, manufacturers, and other business men, which will more and more tend by its increase to keep down exports to the East. The intercourse between the agricultural and manufacturing regions of our country will doubtless increase as fast, and be productive of as much mutual benefit, as any friend of both sections now anticipates; but the home trade within the limits of our North American valley will grow much faster, and possess a vigor as superior to the former as do the great arteries near the heart of those of the limbs of the human system. Western commerce with the Atlantic border is analogous to that of the Eastern and Middle States with Europe.

This trade has had a rapid development, but by no means in proportion to the augmentation of that with their own coast and interior. The foreign commerce of Philadelphia, for instance, is no greater than it was in 1787, when the population of the city and liberties did not exceed 40,000, while its home trade has increased tenfold, and its population become more than five times 40,000. It will probably suprise many of our readers to be informed that the exports and imports of our upper lake region, the past season, have probably exceeded in value those of all the colonies on an average of six years preceding 1775. According to Pitkin, the annual exports from the colonies, of those six years, amounted to £1,752,142, and the imports to £2,732,036. The average annual amount of the exports and imports of this upper lake country for the last three years would be estimated low at $20,000,000. Such are the results of the infantile labors of the young Hercules of the lakes.

The basins of the St. Lawrence and Mississippi constitute nearly all the great interior valley. Each of these basins, when settled to a fair extent, will have a vast commerce of its own; and it will be interesting to ascertain through what channels and through what towns the great intercourse that will naturally grow up between them will be carried on. The time will come, within the present century, when the trade between the northern and southern portions of the North American valley will become more important than that of the whole valley, with the Eastern States and Europe. Until that period arrives, the channels which command most of the Eastern business will be of paramount importance. Let us examine the relative claims of those now used and soon to be prepared for use.

Coming from the East, the first improved communication connecting lake and river trade is the Genesee Valley and Olean Canal. This will compete with the canal from Erie, for the supply of Eastern and European manufactures to much of Western Pennsylvania. In the intercourse between Pittsburgh and the upper lakes, which must soon be of great importance, the channels terminating at Erie and Cleveland will be rivals. To determine which of these is best, requires a more minute knowledge of them than we possess. Supposing them equal, Cleveland being the largest town, and the best mart for such manufactures as Pittsburgh exports, will be sure to attract the greatest portion of this trade.

The Ohio Canal, from Cleveland to Portsmouth, on the Ohio, with its arms to Pittsburgh, to Marietta, and to Athens on the Hocking, furnishes an ample highway for the interchange of productions between the lake regions, and the East and the river regions, embracing Southeastern Ohio, Southwestern Pennsylvania, and Western Virginia. This it holds without having or fearing a rival. How far down in Ohio can its exports from the lakes be carried? This can be ascertained, with some degree of certainty, by comparing it with the Miami Canal route.

The Miami Canal, connecting the lake at Maumee with the Ohio at Cincinnati, embraces at its north end 60 miles of what is known as the Wabash and Erie Canal. It is completed, with the exception of 35 miles, which is to be constructed within the next year. The Eastern trade, by way of the Ohio and Miami Canals, will probably meet on the Ohio, above Maysville. Let us see:

		Miles.
From Lake Erie, at Cleveland,	By Ohio Canal to Portsmouth	306
	By Ohio River down to Maysville	47
	Total	353
From Lake Erie, at Maumee,	By Miami Canal to Cincinnati	235
	By Ohio River up to Maysville	66
	Total	301
Difference in favor of Miami route		52

Sixty miles of the Miami Canal (the Wabash and Erie portion) is more than twice as large as the Ohio Canal. The lockage on the Miami Canal is several hundred feet less than it is on the

Ohio Canal. The conclusion seems unavoidable that the Miami route will send its lake productions and Eastern business as far up the Ohio as Maysville. What will be the limit of its control of this business, South and Southwest? Following the shores of the lakes Westward from Maumee Bay, one will look in vain for any rival channel between the lakes and the Mississippi waters, before reaching the Illinois Canal, at Chicago. The Miami Canal can have no rival in the Eastern business of at least 10,000 square miles of Ohio, the southeastern portion, or 9,000 square miles of Indiana, and nearly the whole of Kentucky. It remains to show where the trade from Lake Erie, by way of the Miami Canal, will probably meet, on equal terms, the same trade by way of the Illinois Canal, on the Mississippi waters; in other words, what portion of the great river valley will be likely to use the one or the other in the transaction of its Eastern business? Will the place at which they may meet on equal terms be at the mouth of the Cumberland river? The Cumberland waters a large extent of fertile country, affords good navigation, and has upon its banks besides many other thriving towns, the important commercial city of Nashville. We will place the distances by the two routes side by side. Lake Erie is the common starting point; for upon her waters must merchandise first come, whether the Erie Canal or St. Lawrence be the channel through which it has been transported:

Lake Erie to the mouth of Cumberland River, by way of Miami Canal.

	Miles.
From Maumee harbor to Cincinnati, by canal	235
" Cincinnati to the mouth of Cumberland, by river	449
Total	684

By way of Illinois Canal.

From Lake Erie to Chicago, by the lakes	750
" Chicago to lower end of Illinois Canal	100
" thence to mouth of Illinois River	267
" thence down the Mississippi to mouth of Ohio	209
" thence up the Ohio to the mouth of Cumberland	57
Total	1,383
Difference in favor of Miami route	699

It will be observed that the Illinois route has an excess of 86 miles of river navigation over the Miami channel, some of which is inferior to that of the lower Ohio. This will, in part, go to balance the excess of canal on the Miami route. The Cumberland Valley, then, clearly belongs to the Eastern rival.

But here comes the more important Tennessee, a river longer than the Rhine, the Elbe, or the Tagus, and navigable into the rich cotton regions of Tennessee, Mississippi, and Alabama.

This is a prize worth contending for. Which of our rival channels will supply its fertile and extensive valley with the large amount of merchandise which its ample means and civilized wants will require? There are but 13 miles separating the mouths of the Cumberland and the Tennessee, so that the Illinois channel gains but 26 miles in comparison with the route just detailed. Still will this route have a balance against it of 673 miles, as compared with its rival, which the following figures will show:

	Miles.
From Lake Erie to mouth of Tennessee, by Chicago	1,370
" " " " Miami and Cincinnati	697
Difference in favor of Miami route	673

We now descend to where the Ohio joins the Father of Waters. Will the trade of the East, through Lake Erie, reach this point? It has already, to some extent, passed out of the Ohio, both up and down the Mississippi, and by a course more circuitous and expensive than either of those I am now comparing, to-wit: that by the Ohio Canal. Let the comparison, then, be made at this point between our rivals. From the mouth of the Tennessee to the Mississippi the distance is 44 miles:

	Miles.
From Lake Erie to mouth of Ohio, by Chicago and St. Louis	1,326
" " " " Maumee and Cincinnati	741
Difference in favor of the latter route	585

In going up the Mississippi, we must, of course, come to the point where the advantages of the two routes will be equal. Is that point at St. Louis?

	Miles
From Lake Erie to St. Louis, by Chicago	1,150
" " " Miami Canal and Ohio and Miss.	917
Difference	233

Thus it appears that St. Louis will have a choice of two nearly equally desirable routes of communication with New York, by way of Lake Erie. Another route from Lake Erie to St. Louis, by way of the Wabash and Erie Canal, would be much better.

	Miles.
From Maumee to Covington, on Wabash, by canal	270
" Covington by proposed rail to St. Louis	196
Whole distance	466

On the whole, it seems to us quite plain that of all the channels of trade now open and being opened in our extensive country, no one of the same extent is destined to be the medium of such extensive commercial operations as the canal which connects, by the shortest route, Lake Erie with Cincinnati.

When the day shall arrive that witnesses the predominance of the home trade of the North American valley over that which is carried on with the Eastern States and with Europe, and the intercourse between the northern and southern portions of it takes the place of that which now is carried on with the old States; and when, also, the shores of the upper lakes shall be brought under cultivation, and become densely settled, the just claims of the Chicago route to participate largely in the trade between the lakes and the central and lower Mississippi Valley will be greatly enlarged. Then she will be the port from which supplies of Southern productions will be drawn for all the borders of the great Lakes Michigan and Superior, and the northern shores of Lakes Huron and Iroquois, and through which will be sent southward most of the surplus productions of those extensive regions. But the Miami Canal, as soon as completed, will fall into possession of a well-peopled and highly-cultivated region of great extent, whose productions will rush through, from both extremes, the moment it is rendered navigable. Not less than two millions of people, living in the southwestern part of Ohio, the southeastern part of Indiana, and almost throughout the entire States of Kentucky and Tennessee, will make it the medium through which their imports from New York will be received; and not less than one million, living on the borders of the lakes, will depend on it for the introduction of sugar, cotton, rice, and other peculiar productions of the South. If the agricultural productions put afloat upon it incline as strongly for a market to the lake end of this as of the Ohio Canal (and we cannot doubt that they will still more so, for it is a better and more direct canal, being 71 miles shorter), then will they pass along its whole line, from south to north, embracing the vast surplus gathered in at Cincinnati. From the lake there will be sent up this canal, besides merchandise, great quantities of pine lumber, building stone (which abounds near its northern termination), mineral coal, salt, gypsum, lake-fish, and doubtless many other articles. It seems clear, then, that, of all the thoroughfares provided for the promotion of trade between the lake and the river valleys of the West, the Miami Canal is to be by far the most important.

But there are rivals in the New York trade with the river valley, which nowhere touch the lakes or the Erie Canal. These are, first, the Philadelphia and Pittsburgh, by canals and railroads; and, second, the Ocean, Gulf, and River route, by way of New Orleans. It remains to compare these with the Miami channel.

The present leading emporium of river commerce, Cincinnati, will be the assumed point of receipt and shipment.

For expense of the carriage of goods (100 pounds) at present rates, from New York to Maumee, 800 miles...	80
Insurance of 100 pounds at one-half of one per cent. on estimated average value of $16...	08
From Maumee to Cincinnati, by Miami Canal, 255 miles...............	45
Amount...	$1 33

By Philadelphia and Pittsburgh from New York, the freight and charges will be—

To Philadelphia, per 100 pounds...	12
" Pittsburgh, " "	$1 10
" Cincinnati, " "	20
" Insurance of 100 pounds at 1¼ per cent. on $16.................	20
Amount..	$1 62

The time required by each will be nearly the same when the Ohio is in good navigable condition. It is, however, well known that the river between Pittsburgh and Cincinnati is not to be relied on for any considerable portion of the season, when the Pennsylvania canals are navigable; and the merchant, who, above all things, desires certainty and expedition in his operations, will hardly decline the reliable and safe route by the lake, in favor of the more uncertain and hazardous one by the Ohio river. For his earliest spring supplies, he will doubtless receive a small stock by the Pennsylvania and Baltimore routes; but for his main supply, he will as certainly adopt the safest and cheapest channel. Which of these routes will be the best for the surplus of agriculture shipped to New York? Contracts by responsible lines have been made for the transportation of flour, from Lafayette, on the Wabash, to New York, for from $1 45 to $1 50 per barrel. The distance from Lafayette to Maumee is 215 miles, 20 miles less than from Cincinnati. We will, therefore, put the cost of sending a barrel of flour —

From Cincinnati to New York, at..		$1 55
Pittsburgh route	Up the Ohio to Pittsburgh..................................	45
	Canal and railroad to Philadelphia....................	$1 10
	Thence to New York..	12
	Total..................................per barrel......	$1 67

The difference in the cost of insurance would ordinarily be 6 or 8 cents in favor of the lake route. On pork and other articles, the proportion of expense would be about the same.

Let a comparison now be instituted between the lake and ocean routes; and, first, in the transport of goods Westward:

From New York by Lake Erie, as before detailed, cost per 100 lbs... $1 33

By ocean and river
{ New York to New Orleans.............................. 25
New Orleans to Cincinnati.............................. 63
Insurance to New Orleans, 2 per cent. on $16... 32
Insurance to Cincinnati   "   "   ... 32 }

Total... $1 52

As most of the goods bought in New York for the Cincinnati market would greatly exceed in value our estimate of $16 per 100 pounds, the interior route will have, in regard to all such, a still greater advantage over that by the ocean, and in proportion to the excess of cost above that sum.

Productions sent for the West, having greater weight and bulk in proportion to their value than merchandise coming the other way, can better afford to pay insurance; and, other things being equal, would incline to the New Orleans outlet as the cheapest. The cost of taking flour to the New York market, from all places on the Ohio below Cincinnati (at which point it will be about equal), will be less this way than by the Miami Canal. But flour taken from the West, through New Orleans, brings less in the great Northern markets than that which goes by the lakes, by more than the ordinary cost of carriage from the mouth of the Ohio to Cincinnati. This is well known to be owing to the great liability to damage in going through a hot climate. As a final market, New Orleans is, in general, very fluctuating and uncertain. These facts assure us that nearly all the surplus flour, within reach of the canals leading from the lakes into the Mississippi Valley, will take the northern road to market. For safety from the bursting of boilers, there is no steam navigation in the States, and perhaps not in the world, equal to that of the lakes. On the ocean the use of salt water, and on the Western rivers the use of muddy water, for the boilers, has probably occasioned a large proportion of the explosions that have so greatly augmented the risk of navigation on the Mississippi waters. The pure water of the lakes has proved eminently favorable to safe steam navigation; and the numerous harbors along the American shore of Lake Erie have lessened the risk, and given it an advantage in that respect over the others — Ontario, perhaps, excepted.

But it may be said that, at no distant day, a large portion of the productions of foreign countries brought into the great Western marts for sale will be imported directly from the regions in which they are produced, and that the assuming of

New York as the great center of supply will fail in regard to these, and thus affect the conclusions heretofore drawn. An examination of the various inlets to this foreign trade will not, however, much vary the results on the routes we have contrasted and compared. Is the St. Lawrence, the route for the European supplies, adopted? The Miami and Illinois Canals will still be the channels for its transport to a great part of the Mississippi Valley. Is the Mississippi the chosen channel for the introduction of what are usually called West India and South American products to the upper lakes? Still are these the only rivals in their transportation. Will the Mississippi challenge a comparison with the St. Lawrence, in our anticipated European trade? Such comparison can only result in the triumph of her northern rival. It would not be difficult to prove that, when the canals now being made around the obstructions to navigation from Montreal to the upper lakes shall be finished, so as to admit sea-going vessels to their ports, freight and insurance between Liverpool and the ports of Cleveland, Maumee, and perhaps Chicago, will be lower than to the port of New Orleans. The distance from England or France, by the St. Lawrence, to the ports of Lake Erie, is less, by more than 1,100 miles, than to New Orleans by the Gulf of Mexico. Of the St. Lawrence route, the distance by river and canal, requiring the aid of steam or horse power, may be about 200 miles; and by the Mississippi, from its mouth to New Orleans, upwards of 100 miles. The advantage possessed by the latter of the saving of tolls can hardly be an offset against the 1,100 miles additional length of voyage. Each route will have some peculiar advantage. The northern will build, man, and own, the shipping employed on it; whereas the southern will depend on ships foreign to her port. The southern will be open all the year; whereas the northern will be barred by ice half the year. The favorable effect upon a trade, of being carried on by a maritime people in their own vessels, from their own ports, is made manifest by contrasting the trade of Boston and Portland with that of Charleston and New Orleans. As New Orleans depends mainly on Northern and European vessels to carry on her coastwise and foreign commerce, the lakes can furnish her with their vessels from the middle of November to the middle of April, a season most favorable for the trade of that port, but of entire idleness to lake vessels that do not seek employment on the open seas of more sunny climes.

NUMBER III.—1843.

The increasing tendency to reside in towns and cities which is manifested by the inhabitants of all countries, as they make progress in the arts and refinements of civilization, is sufficiently obvious to most men who think on the subject. But it is not so apparent, to those whose attention has not been particularly turned to the matter, that the improvements of the last century have so much strengthened that tendency as almost to make it seem like a new principle of society, growing out of the combined agency of steam power and machinery. Mr. Hume, who had as clear apprehension of the relations of the various conditions of society, and the operation of the causes modifying them, as any man of his time, expresses the opinion that no city of antiquity probably ever contained more inhabitants than London, which at the time he wrote, near one hundred years ago, was estimated at 800,000. He thought there were internal and inherent causes to check and stop the growth of the most favorably situated cities when they reached that size. Taking the then existing condition of society as the basis of his reasoning, it seems probable that he judged correctly. Neither the spinning jenny, nor the power loom, nor the steam engine, nor the canal, nor the McAdam road, nor the railway, had then been brought into use; nor had the productive power of the soil, aided by science and art, been, at that time, tasked to its utmost to bring forth human sustenance. Mr. Hume looked with the eye of a philosopher on the past and the present; but, in predicting of the future, his mistakes were nearly as numerous as his vaticinations. To judge of the future by the past may seem safe and philosophic to those who believe not in the certain advance of mankind towards a more perfect condition of nature. So to judge was in accordance with the skeptical mind of Mr. Hume. Let us avoid, so far as we may, his mistake; though to us it seems not practicable to avoid falling into some degree of error of the same sort when we undertake to foretell future conditions and events in a rapidly progressive community.

What has been the effect of the improvements, physical and moral, of the past century, on the growth of towns? and what is likely to be their future effect, aided by other and probably greater improvements, on the growth of towns, during the hundred years to come? We define town to mean any place numbering 2,000 or more inhabitants. It is to Great Britain we are to look for the main evidences of the effects of the labor-saving improvements of the last century. The first canal was commenced in that country by the Duke of Bridgewater, no longer ago than 1760. The invention of the spinning jenny, by Hargreaves, followed seven years after. Not long after this, the spinning frame was contrived by the ingenuity of Arkwright. In 1775, Mr. Crompton produced the machine called the mule, a combination of the two preceding. Some time afterwards, Mr. Cartwright invented the power loom, but it was not until after 1820 that it was brought into general use. The steam engine, the moving power of all this machinery, was so improved by Watt, in 1785, as to entitle him to claim, for all important practical purposes, being its inventor. At the same time that these great inventions were being brought into use, the nation was making rapid progress in the construction of canals and roads, and in the duplication of her agricultural products. Indeed, great part of her works to cheapen and facilitate internal trade, including her canals, her McAdam roads, and her railways, have been constructed within the last thirty years. The effect of these, in building up towns, is exemplified by the following facts: Mr. Slaney, M. P., stated in the House of Commons, in May, 1830, that, "in England, those engaged in manufacturing and mechanical occupations, as compared with the agricultural class, were 6 to 5, in 1801; they were as 8 to 5 in 1821; and 2 to 1 in 1830. In Scotland, the increase had been still more extraordinary. In that country they were as 5 to 6 in 1801; as 9 to 6 in 1821; and, in 1830, as 2 to 1. The increase of the general population for the preceding twenty years had been 30 per cent.; in the manufacturing population it had been 40 per cent.; in Manchester, Liverpool, Coventry, and Birmingham, the increase had been 50 per cent.; in Leeds, it had been 54 per cent.; in Glasgow, it had been 100 per cent." The increase of population in England and Wales, from 1821 to 1831, was 16 per cent. This increase was nearly all absorbed in towns and their suburbs, as the proportion of people engaged in agriculture has decreased decidedly with every census. More scientific modes of culture, and more perfect machines and implements, combined with other causes, have rendered an increased amount of human labor unnecessary in the production of a greatly augmented amount of food. In 1831, but one-third of the people of England were employed in the labors of agriculture. In 1841, very little more than one-fourth were so employed.

In Scotland, seven of the best agricultural counties decreased in population, from 1831 to 1841, from 1 to 5 per cent.; whereas, the counties in which were her principal towns increased during the same period from 15 to 34.8 per cent.; the latter being the increase of the county of Lanark, in which Glasgow is situated. The average increase of all Scotland for those ten years was 11.1 per cent. According to Marshall, the increase of population in England for the ten years preceding 1831, was 30 per cent. in the mining districts, 25½ in the manufacturing, and 19 in the metroplitan (Middlesex county), while in the inland towns and villages it was only 7¾ per cent.

The railways, which now traverse England in every quarter, and bring into near neighborhood its most distant points, have been nearly all constructed since 1830. Their effect, in aid of the other works, in augmenting the present great centers of population, will, obviously, be very considerable; how great remains to be developed by the future. London, with its suburbs, has now about 2,000,000 of inhabitants; but she is probably far below the culminating point of her greatness. The kingdom of which she is the commercial heart doubles its population in forty-two years. It is reasonable, then, to suppose that, within the next fifty years, London and the other great *foci* of human beings in that kingdom will have more than twice their present numbers; for it is proved that nearly the whole increase in England is monopolized by the lage commercial and manufacturing towns, with their suburbs.

Will similar causes produce like effects in the United States? In the States of Massachusetts, New York, Pennsylvania, and Ohio, the improvements of the age operated to some extent on their leading towns from 1830 to 1840. Massachusetts had little benefit from canals, railways, or steam power; but her towns felt the beneficent influence of her labor-saving machinery moved by water power, and her improved agriculture and common roads. The increase of her nine principal towns, commencing with Boston and ending with Cambridge, from 1830 to 1840, was 66,373, equal to 53 per cent.; being more than half the entire increase of the State, which was but 128,000, or less than 21 per cent. The increase, leaving out those towns, was but 11 per cent. Of this 11 per cent., great part, if not all, must have been in the towns not included in our list.

The growth of the towns in the State of New York, during the same period, is mainly due to her canals. That of the fourteen largest, from New York to Seneca, inclusive, was 204,507, or 64½ per cent.; whereas, the increase in the whole State was less than 27 per cent., and of the State, exclusive of these towns, but 19 per cent. Of this, it is certain that nearly all is due to the other towns not in the list of the fourteen largest.

Pennsylvania has canals, railways, and other improvements, that should give a rapid growth to her towns. These works, however, had not time, after their completion, to produce their proper effects before the crash of her monetary system nearly paralyzed every branch of her industry, except agriculture and the coal business. Nine of her largest towns, from Philadelphia to Erie, inclusive, exhibit a gain, from 1830 to 1840, of 84,642, being at the rate of 39½ per cent. This list does not include

Pottsville or any other mining town. The increase of the whole State was but 21⅓ per cent.

Ohio has great natural facilities for trade, in her lake and river coast; the former having become available only since the opening of the Erie canal, in 1826, and that to little purpose before 1830. She has also canals, which have been constructing and coming gradually into use since 1830. These now amount to about 760 miles. For the last five years, she has also constructed an extent of McAdam roads exceeding any other State, and amounting to hundreds of miles. Her railways, which are of small extent, have not been in operation long enough to have produced much effect. From this review of the State, it will not be expected to exhibit as great increase in town population, from 1830 to 1840, as will distinguish it hereafter. The effects of her public improvements, however, will be clearly seen in the following exhibit: Eighteen of her largest towns, and the same number of medium size and average increase, contained, in 1830, 58,310, which had augmented, in 1840, to 138,916; showing an increase of 138 per cent. The increase of the whole State, during the same period, was 62 per cent. The northwest quarter of the State has no towns of any magnitude, and has but begun to be settled. This quarter had but 12,671 inhabitants in 1830, and 92,050 in 1840.

The increase of the twenty largest towns in the United States, from New York to St. Louis, inclusive, from 1830 to 1840, was 55 per cent., while that of the whole country was less than 34 per cent. If the slaveholding States were left out, the result of the calculation would be still more favorable to the towns.

The foregoing facts clearly show the strong tendency of modern improvements to build towns. Our country has just begun its career; but as its progress in population is in a geometrical ratio, and its improvements more rapidly progressive than its population, we are startled at the results to which we are brought by the application of these principles to the century into which our inquiry now leads us.

In 1840, the United States had a population of 17,068,666. Allowing its future increase to be at the rate of 33⅓ per cent., for each succeeding period of ten years, we shall number, in 1940, 303,101,641. Past experience warrants us to expect this great increase. In 1790, our number was 3,927,827. Supposing it to have increased in each decade, in the ratio of 33⅓ per cent., it would, in 1840, have amounted to 16,560,256; being more than half a million less than our actual number, as shown by the census. With 300,000,000, we should have less than 150 to the square mile for our whole territory, and but 220 to the square mile for our organized States and territories. England has 300 to the square mile. It does not, then, seem probable that our progressive increase will be materially checked within the one hundred years under consideration. At the end of that period, Canada will probably number at least 20,000,000. If we suppose the portion of our country, east and south of the Appalachian chain of mountains, known as the Atlantic slope, to possess at that time 40,000,000, or near five times its present number, there will be left 260,000,000 for the great central region between the Appalachian and Rocky Mountains, and between the Gulf of Mexico and Canada, and for the country west of the Rocky Mountains. Allowing the Oregon territory 10,000,000, there will be left 250,000,000 for that portion of the American States lying in the basins of the Mobile, Mississippi, and St. Lawrence. If, to these, we add 20,000,000 for Canada, we have 270,000,000 as the probable number that will inhabit the North American valley at the end of the one hundred years, commencing in 1840. If we suppose one-third, or 90,000,000, of this number to reside in the country as cultivators or artisans, there will be 180,000,000 left for the towns—enough to people 360, each containing half a million. This does not seem so incredible as that the valley of the Nile, scarcely twelve miles broad, should have once, as historians tell us, contained 20,000 cities.

But, lest one hundred years seem too long to be relied on, in a calculation having so many elements, let us see how matters will stand fifty years from 1840, or forty-seven years from this time. The ratio of increase we have adopted cannot be objected to as extravagant for this period. In 1890, according to that ratio, our number will be 72,000,000. Of these, 22,000,000 will be a fair allowance for the Atlantic slope. Of the remaining 50,000,000, 2,000,000 may reside west of the Rocky Mountains, leaving 48,000,000 for the great valley within the States. If, to these, we add 5,000,000 as the population of Canada, we have an aggregate of 53,000,000 for the North American valley. One-third, or say 18,000,000, being set down as farming laborers and rural artisans, there will remain 35,000,000 for the towns, which might be 70 in number, having each half a million souls. It can scarcely be doubted that, within the forty-seven years, our agriculture will be so improved as to require less than one-third to furnish food and raw materials for manufacture for the whole population. Good judges have said that we are not now more than twenty or thirty years behind England in our husbandry. It is certain that we are rapidly adopting her improvements in this branch of industry; and it is not to be doubted that very many new improvements will be brought out, both in Europe and America, which will tend to lessen the labor necessary in the production of food and raw materials.

The tendency to bring to reside in towns all not engaged in agriculture that machinery and improved ways of intercourse have created, has already been illustrated by the example of England and some of our older States. Up to this time, our North American valley has exhibited but few striking evidences of this tendency. Its population is about 10,500,000; but, with the exception of New Orleans, Cincinnati, and Montreal, it has no large towns. As a whole, it has been too sparsely settled to build up many. Too intent on drawing out the resources of our exuberantly rich soil, we have neglected the introduction of those manufactures and mechanic arts that give agricultural productions their chief value by furnishing an accessible market. This mistake, however, is rapidly bringing about its own remedy. In Ohio, the oldest (not in time but in maturity) of our Western States, the arts of manufacture have commenced their appropriate business of building towns. Cincinnati, with its suburbs, has upwards of 50,000 inhabitants; a larger proportion of whom are engaged in manufactures and trades than of either of the sixteen principal towns of the Union except Lowell. The average proportion so engaged in all these towns is 1 to 8.79. In Cincinnati it is 1 to 4.50. Indeed, our interior capital has but two towns (New York and Philadelphia) before her in number of persons engaged in manufactures and trades. Our smaller towns, Dayton, Zanesville, Columbus, and Steubenville, having each about 6,000 inhabitants, have nearly an equal proportion engaged in the same occupation.

These examples are valuable only as indicating the direction to which the industry of our people tends in those portions of the West where population has attained a considerable degree of density. Of the ten and a half millions now inhabiting this valley, little more than half a million live in towns; leaving about ten millions employed in making farms out of the wilds, and producing human food and materials for manufactures. When, in 1890, our number reaches 53,000,000, according to our estimate, there will be but one-third of this number, to-wit: 18,000,000, employed in agriculture and rural trades. Of the increase up to that time (being 42,500,000), 8,000,000 will go into rural occupations, and 34,500,000 into towns. This would people sixty-nine towns with each half a million.

Should we, yielding to the opinion of those who may believe that more than one-third of our people will be required for agriculture and rural trades, make the estimate on the supposition that one-half the population of our valley, forty-seven years hereafter, will live on farms, and in villages below the rank of towns, the account will stand thus: 26,500,000 (being the one-half of 53,000,000 in the valley) will be the amount of the rural population; so that it must receive 16,500,000 in addition to the

10,000,000 it now has. The towns, in the same time, will have an increase of 26,000,000, in addition to the 500,000 now in them. Where will these towns be, and in what proportion will they possess the 26,500,000 inhabitants?

These are interesting questions, and not so impracticable of an approximately correct solution as, at first blush, they may seem.

One of them will be either St. Louis or Alton. Everybody will be ready to admit that. Still more beyond the reach of doubt or cavil is Cincinnati. We might name also Pittsburgh and Louisville; but we trust that our readers, who have followed us through our former articles, are ready to concur in the opinion that the greatest city of the Mississippi basin will be either Cincinnati or the town near the mouth of the Missouri, be it Alton or St. Louis. Within our period of forty-seven years, we have no doubt it will be Cincinnati. She is now in the midst of a population so great and so thriving, and, on the completion of the Miami canal, which will be within two years, she will so monopolize the exchange commerce at that end of the canal between the river and lake regions, that it is not reasonable to expect that she can be overtaken by her Western rival for half a century.

But such has been the influx of settlers within the last few years to the lake region, and so decided has become the tendency of the production of the upper and middle regions of the great valley to seek a market at and through the lakes, that we can no longer withstand the conviction that, even within the short period of forty-seven years, a town will grow up on the lake border greater than Cincinnati. The following facts, it is believed, will force the same conviction to our readers:

The States of Ohio, Indiana, and Illinois, are bordered by both lake and river. All have large river accommodation, but Illinois has it to an unrivaled extent, whereas it has but one lake port.

Now let us see what has been the relative and positive growth of the river region and lake region of these States, from 1830 to 1840. Southern Ohio, including all south of the National road, and the counties north of that road which touch the Ohio river, had, in 1830, 550,000 inhabitants, and in 1840 730,000; showing an increase of 180,000—equal to $33\frac{1}{2}$ per cent. Northern Ohio, in 1830, numbered but 390,000, which in 1840 had increased to 805,000; exhibiting an increase of 415,000, or 105 per cent. In 1830, Southern Ohio had 160,000 more than Northern Ohio; whereas, in 1840, the latter excelled the former 75,000. The preponderance of the lake region has not been owing to the superiority of its soil, or the beauty of its surface; for, in these respects, it is inferior to its Southern rival.

Let us now see how the river and lake regions of Indiana compare, in 1830 and 1840. The National road is the dividing line:

Southern Indiana had in 1830.....252,000
Northern Indiana had in 1830.... 89,000
Southern Indiana had in 1840.....397,000
Northern Indiana had in 1840.....278,000
Southern Indiana, in 1830 ....252,000 } Gain 145,000, or 58 per cent.
    "   "   " 1840 ....397,000 }
Northern Indiana had in 1830..... 89,000 } Showing a gain of 189,000, or
    "   "   " " 1840.....278,000 }   212 per cent.

Such has been the rapidity of settlement of the northern counties of Indiana, for the three years since the census was taken, that we cannot doubt that the north has nearly overtaken, in positive numbers, the south half.

Illinois exhibits the preference given for the lake region in a still more striking manner. A line drawn along the north boundaries of Edgar and Coles counties, and thence direct to the town of Quincy, on the Mississippi, will divide the State into two nearly equal parts. The three counties of Morgan, Sangamon, and Macon, we divide equally, and give two-thirds of Adams to the north and one-third to the south.

Southern Illinois had in 1830......122,732
Northern Illinois had in 1830..... 33,852
Southern Illinois had in 1840.....242,873
Northern Illinois had in 1840.....232,222
Southern Illinois, in 1830 ....122,732 } Showing a gain of 120,141, equal
    "   "   1840..........242,873 } to 97 per cent.
Northern Illinois had in 1830..... 33,852 } Showing a gain of 198,370, equal
    "   "   1840.....232,222 } to 586 per cent.

There can be no doubt with those who know the course of immigration that Northern Illinois, at this time, contains many thousands more than Southern Illinois.

It may be said that the lake region of these States, being of more recent settlement, and having more vacant land, has, chiefly on that account, increased more than the river region. This might account for a *higher ratio*, but it would not account for a greater *amount* of increase. For instance: the State of New York, between 1820 and 1830, had a greater amount of increase than any Western State, though most of them increased in a far higher ratio. So, by the census of 1840, it appears that the *amount* of increase of Ohio for the ten years previous was about three times as great as that of Michigan, although the *ratio* of increase of Michigan was more than nine times as high as that of Ohio.

Let us compare, then, the *amount of increase* of the lake and river regions of these States:

Increase from 1830 to 1840 of
{ Northern Ohio 413,000
" Indiana 189,000
" Illinois .... 198,370 }

800,370

Increase from 1830 to 1840 of
{ Southern Ohio 180,000
" Indiana 145,000
" Illinois .... 120,141 }

445,141

Arkansas and Michigan, were it not that the latter has the advantage of not holding slaves, would afford almost a perfect illustration of the preference given to the lake region over the river country. Each has extraordinary advantages of navigation of its peculiar kind. No State in the valley has as extensive river navigation as Arkansas, and no State can claim to rival Michigan in extent of navigable lake coast.

In 1830, Michigan had a population of 32,538
" Arkansas " " 30,388
In 1840, Michigan numbered 212,276
" Arkansas " 97,578

These facts exhibit the difference in favor of the lake country sufficient to satisfy the candid inquirer that there must be potent causes in operation to produce such results. Some of these causes are apparent, and others have been little understood or appreciated. The staple exports, wheat and flour, have for years so notoriously found their best markets at the lake towns, that every cultivator, who reasons at all, has come to know the advantage of having his farm as near as possible to lake navigation. This has, for some years past, brought immigrants to the lake country from the river region of these States, and from the States of Pennsylvania, Maryland, and Virginia, which formerly sent their immigrants mostly to the river borders. The river region, too, not being able to compete with its northern neighbor in the production of wheat, and being well adapted to the growth of stock, has of late gone more into this department of husbandry. This business, in some portions, almost brings the inhabitants to a purely pastoral state of society, in which large bodies of land are of necessity used by a small number of inhabitants. These causes are obviously calculated to give a dense population to the lake country, and a comparatively sparse settlement to the river country. There are other causes not so obvious, but not less potent or enduring. Of these, the superior accessibility of the lake country from the great northern hives of emigration, New England and New York, is first deserving attention. By means of the Erie canal to Oswego and Buffalo, and the railway from Boston to Buffalo, with its radiating branches, these States are brought within a few hours' ride of our great central lake; and at an expense of time and money so small as to offer but slight impediment to the removal of home and household gods. The lakes, too, are about being traversed by a class of vessels, to be propelled by steam and wind, called Ericsson propellers, which will carry immigrants with certainty and safety, and at greatly reduced expense.

European emigration hither, which first was counted by its annual thousands, then by its tens of thousands, has at length swelled to its hundred thousands, in the ports of New York and Quebec. These are both but appropriate doors to the lake country. It is clear, then, that the lake portion will be more populous than the river division of the great valley. This is one reason why the former should build up and sustain larger towns than the latter.

It has been proved that an extensive and increasing portion of the river region seeks an outlet for its surplus productions through the lakes. In addition to the proof given on that subject, we will compare the exports of breadstuffs and provisions of New Orleans and Cleveland—the former for the year beginning 1st of September, 1841, and ending 31st August, 1842; and the latter for the season of canal navigation, in 1842. All the receipts of Cleveland, by canal, are estimated as exports, as there is no doubt that she receives, coastwise and by wagon, more than enough to feed her people. The exports from New Orleans of the enumerated articles, and their price, are as stated in No. 4, vol. 7, of this magazine. Of the articles, then, of flour, pork, bacon, lard, beef, whisky, corn, and wheat—

New Orleans exported to the value of..................................$4,446,989
Cleveland   "   "   4,431,739

The other articles of breadstuffs and provisions received at New Orleans during that year from the interior are of small amount, and obviously not sufficient for the consumption of the city. Not so with Cleveland. The other articles of grain and provisions, shipped last year from this port, added to the above, will throw the balance decidedly in her favor. If we suppose, what can not but be true, that all the other ports of the upper lakes sent eastward as much as Cleveland, we have the startling fact that the lake country, but yesterday brought under our notice, already sends abroad more than twice the amount of human food that is shipped from the great exporting city of New Orleans, the once-vaunted sole outlet of the Mississippi valley. Another striking fact, in favor of the position that on the lakes are to be the leading commercial cities of our valley, is the growth of Cleveland, compared with Portsmouth. When the Ohio canal was completed, that portion of the State traversed by it, lying nearest to Portsmouth, was superior in population and productiveness to that which was nearest Cleveland. Portsmouth is at the river end of the canal, and Cleveland at the lake end:

Portsmouth, including the township in which it is situated, numbered, in 1830..1,464
In 1840...1,844

Increase of Portsmouth, including the township, in ten years........ 380
Cleveland village numbered, in 1830..1,076
" city, including Ohio* City, in 1840................................7,648

Increase of Cleveland in ten years...7,572

The case of Alton and Chicago is calculated to illustrate the same position. The former is so finely situated on the Mississippi, just above the entrance of the turbulent Missouri, at the best point for concentrating the river trade on all sides, and doing the business of one of the finest and best settled portions of Illinois, that we have thought it might yet excel St. Louis, and perhaps rival Cincinnati. The country in its rear was settled long before that lying back of Chicago, and Alton, in consequence, sooner became an important commercial point. How many inhabitants it had in 1830, we have at hand no means of ascertaining. Certain it is that, at that time, it was far more populous than Chicago:

In 1840, Alton numbered...2,340
" Chicago " 4,470

Two short canals—one of about one hundred miles, connecting the Illinois canal with the Mississippi, at or near the mouth of Rock river; and the other of about one hundred and seventy-five miles, connecting the southern termination of the Wabash and Erie canal, at Terre Haute, with the Mississippi at Alton—would, with the canals already finished or in progress, secure to the lakes not less, probably, than three-fourths of all the external trade of the river valley. With the Wabash and Erie, and the Miami canal brought fairly into operation, the lakes will make a heavy draft on the trade of the river valley; and every canal, and railroad, and good highway, carried from the lakes, or lake improvements, into that valley, will add to the draft. The lake towns will then not only have a denser population in the region immediately about them, and monopolize all the trade of that region, but they will have at

---

* Ohio City is separated from Cleveland only by a narrow stream, and has grown since 1830.

least half the trade of the river region. They will be nearer and more accessible to the great marts of trade and commerce of the old States and the old world, and this advantage will be growing, in consequence of the progressive removal of impediments to navigation between the lakes and the ocean.

The facts we have adduced, taken altogether, seem conclusive in favor of the lake towns. As a body, they come out of the investigation decidedly triumphant. But how shall we decide on their relative merits? There are several whose citizens would claim pre-eminence for each—Oswego, Buffalo, Cleveland, the Maumee town (be it Maumee City or Toledo), Detroit, and Chicago. Unless we have failed in our opening article, New Orleans, Montreal, and Quebec, although destined greatly to increase in size and wealth, may be left out of the contest.

Oswego has a fine position as a point of shipment between the lakes and the Eastern States; and, on the completion of the enlarged Welland canal, she will probably gain rapidly on Buffalo in amount of goods forwarded West and produce of the lakes sent to the Hudson. Her water-power will enable her to compete successfully with Rochester in the manufacture of flour, and it must, before many years, be used extensively in other manufactures. As a point for the wholesale or jobbing of goods, she will be inferior to Buffalo. But both towns are too near and too convenient to New York and Boston to become great marts for the sale of European and Eastern manufactures. Buffalo, in her suburb of Black Rock, has an almost exhaustless water-power, which, long within the period of forty-seven years, will make her a considerable manufacturing town. If the Erie canal enlargement should be delayed many years after the completion of the Welland canal, it would not surprise us to see Oswego overtake Buffalo in size and business.

Buffalo has a cramped harbor, and, like Oswego, she has but a small country in her rear to sustain her trade. Her position for carrying on foreign trade, after the enlargement of the Welland canal, will be less favorable than Cleveland, Maumee, Detroit, or Chicago. But, before entering on the comparison of Buffalo and Cleveland, it will be well to lay down some principles that may be reasonably supposed to control or influence their future growth. And first, it may be asserted that a position favorable to an interchange of productions of a large country lying about it, is more advantageous than a situation which merely favors the passage of a great amount of productions through it. Boston and Charleston will illustrate this principle. The former exchanges, in her own market, the productions gathered into it from the coast, from the interior, and from foreign countries. Charleston is far less a gathering point of commodities, but has a much larger value passing through the hands of her merchants:

7

Boston, between 1830 and 1840, *increased*.................................. 33,611
Charleston, "   "   "   *decreased*.................................. 1,628

Other causes, no doubt, aided in this result; but that under consideration we believe to have been the chief.

Second. While a country is new, the first exchanges will be of agricultural products of one climate for those of a different climate, and of agricultural products for manufactured articles of first necessity. As society progresses in wealth, in addition to these articles, finer fabrics and of greater variety become the subject of exchange; so that when its condition approximates that of England, much of its exchangeable capital comes to be composed of the highly wrought productions of the various cities—each mainly engaged in its own peculiar production, and therefore dependent on all the others for all its articles of consumption, except the one article of its own fabrication.

Let us apply these principles. Buffalo has the advantage of a greater transit of produce and goods. In the former, however, she is not very much in advance, and Cleveland is rapidly gaining upon her. In proportion to her population, Cleveland is already far ahead. As to goods passing to the upper lakes from the old States and Europe, Buffalo will divide chiefly with Oswego the advantages of their receipt and shipment up the lakes. Hers, for some time to come, will be the lion's share—at least until the completion of the Canadian improvements. But these goods, though of great value, will employ no great amount of tonnage, especially when sugar, molasses, cotton, rice, and tobacco, shall be sent to the lakes by the Miami and Illinois canals, as will soon be the case.

Long within the period under consideration, the position of Cleveland will be much more favorable for concentrating the business of the surrounding country than that of Buffalo. Canada will, before that time, form a part of our commercial community, whether she be associated with us in the government or not. She will then have about five millions of people. The American shores of the lakes lying above the latitude of Cleveland will be still more populous.

Cleveland is the lake port for the great manufacturing hive at the head of the Ohio river—so made by the Mahoning canal, which connects her with Pittsburgh. She commands, and she will long command, by means of her five hundred miles of canal and slack-water navigation, the trade of a part of Western Pennsylvania, most of Western Virginia, and nearly all the east half of the State of Ohio, in the intercourse of their inhabitants with the lake coasts, the Eastern States, Canada, and Europe. Her position is handsome; and, although her water-power is small, the low price of coal will enable her to sustain herself as a respectable manufacturing town. Her harbor, like that of Buffalo, though easy of entrance, is not sufficiently capacious. If coal should not be found on Lake Huron, more accessible to navigation than the beds on the canal south of Cleveland, this article will greatly increase her trade with the other lake ports. It is now sold on her wharves at eight cents per bushel.

A glance at a map of the country will suffice to show that Buffalo is not well situated to be a place for the exchange of agricultural productions of the cold regions for those of the warm regions of the valley. In that respect Cleveland, though not unrivaled, is clearly in a better position than Buffalo. As a point for exchanging the products of the field for manufactured goods, Buffalo will not probably for any long time have the advantage of Cleveland. Such traders as live within the influence of the canals and rivers that pour their surplus products into Cleveland, and stop short of New York and Boston, will, it seems to us, be more likely to purchase in Cleveland than in Buffalo. Not every man who supplies a neighborhood with store-goods relishes a voyage on the sometimes tempest-tossed waters of the lake; and, as we before remarked, Buffalo now being but a few hours' ride from New York and Boston, by a pleasant and safe conveyance, will hardly stop many purchasers of goods from those great markets. On the completion of the Canadian canals, Cleveland will have the advantage of Buffalo in foreign trade, for the following reasons: Her articles of export will be cheaper, and by that time, as we believe, more abundant. By means of her canals and roads Cleveland is a primary gathering-point of these articles. Not so Buffalo. To arrive at her storehouses, these products must be shipped from the storehouses of other ports up the lakes, where they must be presumed to bear nearly the same price as at Cleveland. The cost of this shipment, together with a profit on it, will then be added; and, by so much, enhance their price in Buffalo. A vessel entering Lake Erie by the Welland canal, seeking a cargo for a foreign port, would therefore clearly prefer going to the head of the market, where it could be bought at the cheapest rate. If the difference in price of exportable products, between the market at Buffalo and the maket at Cleveland, is such as to warrant the payment of a freight to Buffalo, and the cost of a transhipment there to the foreign vessel, there can be no doubt of its being the interest of the foreign vessel to proceed directly to Cleveland for her cargo; and so to any other considerable market on Lake Erie, and probably the lakes above. It seems likely, therefore, that within our allotted period of forty-seven years Cleveland will be larger than Buffalo or Oswego.

Is it probable that, within the period under consideration, Cleveland will have a successful rival in Maumee, Detroit, or

Chicago? It will be proper, on account of its comparative obscurity and the peculiarity of its position, for us to explain in regard to Maumee.

The estuary of the Maumee river receives the tide of Lake Erie, and the waters of the river, at a point thirteen miles above its mouth. This estuary forms a harbor of Lake Erie, thirteen miles long, with a navigable channel of about one hundred rods. Its depth, in a low stage of the lake, is from six and a half to twenty-four feet. It is entered by a wide channel through the bay, having in its shoalest part 8.25 feet when the lake is in its lowest stage. On the southwest end of this harbor Maumee City and Perrysburg are situated, the former on the north and the latter on the south bank. Both are on the same plane, sixty-three feet above the harbor. Eight miles below, on the north bank, is Toledo, most of it on a plane about forty-five feet high; and three or four miles below Toledo is Manhattan, elevated in its highest part about twenty-five feet above the water. Their population, respectively, including the civil township, was, according to the census of 1840—Maumee City, 1,290; Perrysburg, 1,065; Toledo, 2,053; Manhattan, 282. Each of these places has access to the canal by a side-cut and flight of locks. It is not our purpose to decide on their relative merits; but for convenience, and because it is the name of the harbor, we will call the successful point *Maumee*.

The contest is now fairly narrowed down to Cleveland, Maumee, Detroit, and Chicago. Which of these will be greatest in 1890? We have shown in a previous article (No. 2 of this series) that the Miami canal route will command the Eastern and European trade of Kentucky, most of Tennessee, large portions of Ohio, Indiana, and Illinois, and small portions of Missouri, Arkansas, Mississippi, and Alabama. So long, then, as this Eastern and European trade shall continue of paramount importance to the great country embraced by the description above, as controlled by the Miami canal, so long must the point most favorably situated at its lake termination have the advantage of the other lake towns. We have also shown, in the same article, that the interior exchanges, the exclusive home-trade of the North American valley, between the lake regions of the north and the river regions of the south, will be chiefly carried on through the same Miami canal. Of the towns now under comparison, Maumee is the smallest and Detroit the largest. This, in the minds of the superficial, will be taken as conclusive in favor of the latter. The claim, in favor of a town just emerging from the forest to rival, at a future time, an already populous city, is usually met by ridicule from such persons; and, in general, is treated with little attention or respect by any class. We dare say that when the people of the city of old and renowned York were informed that, in the wilds of America, some settlers had named their collection of rude houses New York, they felt no other emotion than contempt, and treated the presumptuous ambition of the settlers with derision. It is probable that the settlers of old Boston held in like contempt the assumption of the name of their town by those who planted the capital of New England. Who, forty-seven years ago, would not have ridiculed the opinion, if any one had been visionary enough to express it, that, within that time, there would grow up in the valley of the Ohio a city containing fifty thousand inhabitants; and that, within the same period, that part of the Northwestern territory, now composing the State of Ohio, would contain nearly two millions of people? We then had, as a basis of increase, but four millions; whereas it is now over eighteen millions—and, including Canada, near twenty millions. For the past forty-seven years, our growth has been from four millions to near twenty millions. During the next forty-seven years it will be, according to our estimate, from near twenty millions to seventy-seven millions; or, according to the more elaborate and probably more correct estimate of Professor Tucker, fifty-five millions. This increase will certainly make it necessary that many towns, now small, should become great; and sensible men, when contemplating their probable destiny for half a century in advance, will look at the natural and artificial advantages of our lake towns, rather than at the few thousands, more or less, of the present population. The towns under consideration are all destined to be large. The leading advantages of Cleveland have been already stated. Detroit has a pleasant site and a noble harbor. A few McAdam roads, leading north, northwest, and west, into the interior, would give her the direct trade of a large and fertile portion of Michigan. Until such roads, or a reasonably good substitute, are made, the railways leading north and west will, at least while they are new and in good order, make the chief gathering points of trade at their interior terminations and at convenient points on their line. Pontiac, Ypsilanti, Ann Arbor, and other towns west, will cut off from Detroit, and center in themselves the direct trade with the farmers, which, with good wagon roads, without the railways, would have centered in Detroit. One train of cars will now bring to her warehouses what would have been brought to her stores by one hundred wagons. These wagons would have carried back store-goods and the products of Detroit mechanics, whereas these will now be bought in the interior towns. Most of the money borrowed by Michigan, and for which she is so largely in debt, has been expended with a view to center the trade of the State mainly in Detroit and Monroe; but we much doubt whether the effect of the railways constructed for that purpose will not be the reverse of what was anticipated by their projectors. The effect of the Erie and Kalamazoo railway, from Toledo to Adrian, has been to convert a small cluster of houses at the latter place into a flourishing town of near two thousand inhabitants; while at Toledo its effect has been mainly perceptible in the filling a few warehouses with produce and goods, and leaving its business street nearly deserted of wagons, and its hotels almost destitute of any but minute-men travelers. We do not believe that machines so expensive and so complicated in their construction and operation as railways can be sustained in an agricultural country so new and sparsely settled as Michigan. But whether this is a correct view or not matters little to Detroit, if, as we suppose, her railways will but substitute trains of cars, passing through to her warehouses, for the throng of wagons that, but for her railways, would have crowded her broad avenue. The extent of country that will find in Detroit its most convenient point of exchanges is not very great, yet sufficient when well settled and improved to sustain her in a considerable advance beyond her present size and business.

If we now narrow down our comparison by leaving out Detroit, we trust we shall be justified by our impartial readers.

Cleveland, Maumee, and Chicago, only remain to contest the prize. Of these, Maumee alone has a harbor capacious enough to accommodate the commerce of a great city. Good harbors may be made, without a very heavy cost, at Cleveland and Chicago, either by excavating the low grounds bordering their present harbors, or by break-waters and piers in the lakes outside. Some expenditure will also be needed to deepen the entrance into Maumee harbor and to remove obstructions within it. In water-power Maumee has greatly the advantage over her rivals. Chicago has and she can have none. Cleveland has but a small amount; whereas Maumee has it to an extent unrivaled by any town on the lake borders, above Buffalo—and it is so placed as to possess the utmost availability. Along her harbor for thirteen miles the canal passes on the margin of the high bank that overlooks it. This canal—a magnificent mill-race, averaging near seven feet deep, and seventy feet wide at the water line—is fed from the Maumee river, seventeen miles above the head of the harbor, and is carried down on the level of low water in the river above, for twenty-two miles, to a point two miles below the head of the harbor, where it stands on a table land, sixty-three feet above the harbor. Descending, then, by a lock, seven feet, the next level is two miles long, and stands fifty-six feet above the harbor. Descending again, by a lock, seven feet, the level below is three and a half miles long, and stands forty-nine feet above the harbor. Again descending, within the city of Toledo, by four locks, thirty-four feet, the next and last level is nearly five miles long, and stands fifteen feet above the harbor

At many points of these thirteen miles, the water may be used conveniently from the canal to the harbor; and, at most of these points, it may be used directly on the harbor. The Board of Public Works, in their last report, say: "From the experience the Board have had as to the quantity of water required to propel one pair of four and a half feet mill-stones, with all the labor-saving machinery necessary for the manufacture of superfine flour, they are fully of opinion that there will be power sufficient, that can be used on these levels, to propel two hundred and twenty-five pairs of stone." The lowest estimate for the dryest season allows it this amount of power. At other times, the amount is so great that, for all practicable purposes for many years to come, it may be set down as without limit. The current occasioned by the use of the great power estimated by the Board would not be one mile an hour. If more should be used, so as to occasion a current of one mile and a half an hour, the obstruction to navigation would be rather nominal than real. The down-freights for many years will be three or four times as heavy as the up-freights. The current, then, would aid the movement of three or four tons where it would hinder the movement of one ton. If, at some future day, the water furnished during the dry seasons should not be sufficient for the machinery then needed at this point, steam may be used temporarily during the lowest stage of water. Coal will be afforded at ten cents per bushel; and wood, for many years, will not cost more than $1 50 to $2 00 per cord. Will this be a good point for the use of water-power? This will depend on its facilities for procuring raw materials and distributing the manufactured articles to consumers. As to facilities for procuring wheat for the manufacture of flour, there can be, as all will admit who know the country within reach of the canals, no better point in the States. Sheep are so rapidly multiplying in Indiana and Illinois, and are already so abundant in the Miami country of Ohio, that a supply of wool to an extent beyond any probable demand for its manufacture may be safely anticipated. As to cotton, it has been proved that the Miami canal is the best channel for its import to the lakes. From Florence, in Alabama, it may be brought to the factory on the Maumee by a course three hundred miles shorter than its usual route to New Orleans. Should the Tennessee river fail to furnish enough cotton, the Arkansas, and the Mississippi above the mouth of the Arkansas, will be able to supply any additional demand. For the distribution of the manufactured goods, the whole West is easily accessible by means of lakes, canals, and rivers.

As a point for manufacturers and mechanics, the aids and facilities above mentioned give Maumee an incontestable superiority over Cleveland and Chicago. Let us now compare their commercial advantages. Those of Cleveland have been already set forth to some extent, comparing her claims with those of Buffalo. In the exchange of agricultural products of a warm and of a cold climate, Cleveland, by her canals and her connection with the Ohio, can claim south, as against the Miami canal, no farther than Western Virginia and Eastern Kentucky. Maumee will supply the towns on the Lakes Erie, Huron, and probably Ontario, with cotton, sugar, molasses, rum (may its quantity be small), rice, tobacco, hemp (perhaps), oranges, lemons, figs, and, at some future day, such naval stores as come from the pitch-pine regions of Tennessee, Mississippi, and Louisiana. Chicago will furnish a supply of the same articles to Lake Michigan, Lake Superior, when that lake becomes accessible to her navigation, and perhaps the northern portion of Lake Huron. How important these commodities are in modern commerce need not be enlarged on in a magazine whose readers are mostly intelligent merchants. During the forty-seven years under consideration, the countries to be supplied with these articles from Maumee will continue to be more populous than those depending on Chicago for their supply. This position seems too obvious to need proof. It is clear, then, that as a point of exchange of agricultural products of different climates, Maumee has advantages over Chicago—the only place on the lakes that can set up any pretensions of rivalry in this branch of trade.

What are the relative merits of these towns for the exchange of agricultural products for the manufactures of Europe and the Eastern States? The claims of Cleveland, in this respect, have already been considered; and to some extent, also, those of Maumee. The control of Cleveland, south and southeast, embraces a country of about 40,000 square miles; being a quarter larger than Ireland. For early spring supplies, and light goods, this domain may be invaded from Philadelphia and Baltimore; but for the shipment east, and the bulk of goods from New York and Europe, it belongs legitimately to Cleveland.

Maumee will have in this trade the chief control of not less than 100,000 square miles—say 12,000 in Ohio, 30,000 in Kentucky, 30,000 in Indiana, 10,000 in Illinois, 13,000 in Tennessee, 5,000 in Mississippi and Alabama, and 5,000 in Michigan—to say nothing of her claims on small portions of Missouri and Arkansas. This domain is half as large as the kingdom of France and twice as fertile. The Miami canal, connecting Maumee with Cincinnati, will, with that part of the Wabash and Erie which forms the common trunk after their junction, be two hundred and thirty-five miles long. The Wabash and Erie canal, from Maumee to Terre Haute, will be three hundred miles long. Of this, all but thirty-six miles, at its northern extremity, will be in operation the present season. By means of these canals, and the rivers with which they communicate, great part of this extensive region will enjoy the advantage of a cheap water transport for its rapidly increasing surplus.

Chicago, on the completion of the Illinois canal, may command, in its exchange of agricultural for manufactured products, an extent of territory as large as that controlled by Maumee. Admitting it to be larger, and of this our readers must judge for themselves, it does not seem to us probable that within the forty-seven years it can even approximate in population or wealth to the comparatively old and well-peopled territory that comes within the range of the commercial influence of Maumee. We have not sufficient data on which to calculate the extent of country that will come under the future commercial power of Chicago. That it is to be very great seems probable from the fine position of that port in reference to the lake, and an almost interminable country southwest, west, and northwest of it. An extension of the Illinois canal to the mouth of Rock river seems destined to give her the control of the Eastern trade throughout the whole extent of the upper Mississippi, except what she now has by means of the Illinois river. She will also probably participate with Maumee in the lake trade with the Missouri river and St. Louis. On the whole, we deem Chicago alone, of all the lake towns, entitled to dispute future pre-eminence with Maumee. The time may come, after the period under consideration, when the extent and high improvement of the country making Chicago its mart for commercial operations, may enable it at least to sustain the second place among the great towns of the North American valley, if not to dispute pre-eminence with the first.

When we properly consider the future populousness of our great valley, the tendency of modern improvements to build up large towns, the great and increasing inclination of population and trade to and through the lakes, and the decided advantages which Maumee possesses over any other lake port, we need not fear being over sanguine in anticipating for the leading town on that port a growth unrivaled by any city whose history has been recorded.

The conclusions to which we have come, in this and the preceding articles on internal trade, are not expected to be universally or generally acceptable. Many of them run counter to the hopes and preconceived opinions of too many persons for us to expect that they will be considered with candor, or judged with impartiality. The facts therein contained will be encountered with less alacrity. On these we rely. For these we ask a dispassionate and fair examination. If other and different conclusions are deducible from them than those we have drawn, it would give us pleasure to acknowledge our error and correct it. But if, after a thorough examination of the subject, we have gone beyond the anticipations of men who, with more ability, have bestowed much less thought on it, let them not condemn merely because our conclusions seem to them extravagant; but let them examine for themselves, or, if they will not do that, let them hesitate before they pass a hasty judgment on what we have investigated with the utmost care, and with an earnest desire to arrive at the truth. J. W. S.

Number IV.—1848.

COMMERCIAL CITIES AND TOWNS OF THE UNITED STATES.

OUR CITIES — ATLANTIC AND INTERIOR.

All people take pride in their cities. In them naturally concentrate the great minds and the wealth of the nation. There the arts that adorn life are cultivated, and from them flows out the knowledge that gives its current of thought to the national mind.

The United States, until recently, have had large cities in the hope rather than in the reality. It is but a few years since our largest city reached a population of one hundred thousand. Long before that period sagacious men saw, in the rapid growth of the country and the aptitude of our people for commerce, that such positions as those occupied by Philadelphia and New York must rapidly grow up to be great cities. This, however, was by no means the common belief in this country; and our transatlantic brethren treated with undisguised ridicule the idea that these places could even rival in magnitude the leading cities of their own countries. New York is now sometimes called the London of America. Not that those calling her so suppose she will ever come up to that mammoth in size and importance, but because she holds in the New World the relative rank which London holds on the Old Continent.

It is believed that few persons, at this time, have a sufficiently high appreciation of the future grandeur of New York; and yet fewer can be found who doubt that she will always continue to be the commercial capital of America. If this should be her destiny, the imagination could hardly set a limit to her future growth and grandeur. It would be presumptuous to say that her population might not reach five millions within the next century and a half. Of the few persons who have doubted her continual supremacy, most have given the benefit of the doubt to New Orleans. This outport of the great central valley of North America was believed to command a destiny, when this valley should become well peopled, that might eclipse the island city of the Hudson.

Some twenty years ago, the writer, then living in a southeastern State, was convinced that the greatest city must, in the nature of things, at a not very distant day, grow up in the interior of the continent. Of this opinion he thinks he was the *inventor*, and, for many years, the *sole proprietor*. If it had been the subject of a patent, no one would have been found to dispute his claim to the exclusive right to make and vend (if that could be said to be vendible which no one would be prevailed on to take as a gift). That such an opinion should appear absurd and ridiculous may very well be credited by most people, who consider it not much less so now. The largest city of the interior was then Cincinnati, having scarcely 20,000 inhabitants; and the sum total of all the towns in the great valley scarcely exceeded 50,000. St. Louis at that time had but 5,000, and Buffalo about the same number. Here, then, was a basis very small for so large an anticipation. Who could believe that St. Louis, with 5,000 people, could possibly, within the short period of 150 years, become greater than New York, with a population of near 200,000? But what seemed most ridiculous of all was that the future rival of the great commercial emporium should be placed a thousand miles from the ocean, where neither a ship of war nor a Liverpool packet could ever be expected to arrive.

Since 1828, some changes of magnitude have taken place; and the writer's *exclusive right* might now be questioned. There are now other men, considered sane men, who believe the great city of the nation is to be west of the mountains, and quite away from the salt sea. Governor Bebb, in a late address before the Young Men's Library Association of Cincinnati, expressed his decided belief that Cincinnati would, in the course of a century, become "the greatest agricultural, manufacturing, and commercial emporium on the continent." There are other men now, not much less distinguished for knowledge and forecast than Governor Bebb, who entertain the same belief. What has wrought this change of opinion? Time, whose business is to unfold truth and expose error, has given proofs which can no longer be blinked. The interior towns have commenced a growth so gigantic that men must believe there is a power of corresponding magnitude urging them forward — a power yet in its infancy, but unfolding its energies with astonishing rapidity.

Let us make some comparisons of the leading Eastern and Western cities. New York was commenced nearly 200 years before it increased to 100,000 people. Cincinnati, according to Governor Bebb, has now, fifty years from its commencement, 100,000 inhabitants. Boston was 200 years in acquiring its first 50,000. New York, since 1790, when it numbered 33,131, has had an average duplication every fifteen years. This would make her population in 1850, 530,096. This is very near what it will be, including her suburb, Brooklyn.

Cincinnati has, on the average, since 1800, when it had 750, doubled her numbers every seven years.

NEW YORK.

1790	33,131	1820	132,524	1850	530,096
1805	66,262	1835	265,048		

CINCINNATI.

1800	750	1821	6,000	1842	48,000
1807	1,500	1828	12,000	1849	96,000
1814	3,000	1835	24,000		

It appears from this table that, on the average of fifty years, Cincinnati, the leading interior town, has doubled her population every seven years; while New York, on the average of sixty years, has scarcely doubled hers in every period of fifteen years. If New York is compared to Cincinnati during the same fifty years, it will be seen that the period of her duplication averages over fifteen years. She had, in 1800, 60,489. Doubling this every fifteen years, she should have, in 1850, nearly 650,000. This number will exceed her actual population more than 100,000, whereas Cincinnati in 1850 will certainly exceed 96,000.

Let us now suppose that, for the next fifty years after 1850, the ratio of increase of New York will be such as to make a duplication every eighteen years, and that of Cincinnati every ten years. New York will commence with about 500,000, which will increase by the year

1868 to..........1,000,000 | 1886 to..........2,000,000 | 1904 to..........4,000,000

Cincinnati will commence in 1850 with at least 100,000, which will double every ten years; so that in

1860 it will be..	200,000	1880 it will be..	800,000	1900 it will be..	3,200,000
1870 "	400,000	1890 "	1,600,000	1904 "	4,066,667

The resulting figures look very large, and, to most readers, will appear extravagant.

Let us suppose the duplication of New York, for the next 100 years, to be effected on an average of twenty years, and that of Cincinnati of twelve years.

CHANGE OF NATIONAL EMPIRE.   117

NEW YORK IN

| 1850 | 500,000 | 1890 | 2,000,000 | 1930 | 8,000,000 |
| 1870 | 1,000,000 | 1910 | 4,000,000 | 1950 | 16,000,000 |

CINCINNATI IN

1850	100,000	1886	800,000	1922	6,400.000
1862	200,000	1898	1,600,000	1934	12,800.000
1874	400,000	1910	3,200,000	1946	25,600,000

This looks like carrying the argument to absurdity; but if these two leading cities be allowed to represent all the cities in their sections respectively, the result of the calculation is not unreasonable. It is not beyond possibility, and is not even improbable.

The growth of the leading interior marts, since 1840, has been about equal to the average growth of Cincinnati for fifty years past. This growth, for the last eight years, according to the best information to be obtained, has been more than 115 per cent., as the following table will show:

	1840.	1848.		1840.	1848.
Cincinnati	46,900	95,000	Detroit	9,000	17,000
St. Louis	16,000	45,000	Milwaukee	2,000	15,000
Louisville	21,000	40,000	Chicago	5,000	17,000
Buffalo	18,000	42,000	Oswego	5,000	11,000
Pittsburgh	31,000	58,000	Rochester	20,000	30,000
Cleveland	6,000	14,000			
Columbus	6,000	14,000	Total	191,000	412,000
Dayton	6,000	14,000			

The growth of the exterior cities for the same period has been about 38 per cent., according to the following figures:

	1840.	1848.		1840.	1848.
New York	312,000	425,000	Savannah	11,000	14,000
Philadelphia	228,000	350,000	Mobile	12,000	12,000
Baltimore	102,000	140,000	Brooklyn	36,000	72,000
New Orleans	102,000	102,000	Portland	15,000	24,000
Boston	93,000	130,000			
Charleston	29,000	31,000	Total	940,000	1,300,000

The census for 1840 is our authority for that year. For 1848, we have late enumerations of most of the cities. The others we estimate.

There are doubtless a few inaccuracies in the detail, but not enough to vary the result in any important degree.

In the aggregate our interior cities, depending for their growth on internal trade and home manufacture, increase three times as fast as the exterior cities, which carry on nearly all the foreign commerce of the country, and monopolize the home commerce of the Atlantic coast. This is a fact of significance. It proves that our fertile fields, after supplying food to everybody in foreign lands who will buy, and feeding the cities and towns of the Atlantic States, have sufficed to feed a rapidly growing town population at home. It proves, also, that the Western people are not disposed to accept the destiny kindly offered them by their Eastern brethren, of confining themselves to the handwork of agriculture — leaving to the old States the whole field of machine labor. Although the land on which the people of the great valley have but recently entered is new, the civil, social, and economical condition of this people is advanced nearly to the highest point of the oldest communities. The contriving brain and the skillful hand are here in their maturity. The raw materials necessary to the artisan and the manufacturer, in the production of whatever ministers to comfort and elegance, are here. The bulkiness of food and raw materials makes it the interest of the artisan and manufacturer to locate himself near the place of their production. It is this interest, constantly operating, which peoples our Western towns and cities with emigrants from the Eastern States and Europe. When food and raw materials for manufacture are no longer cheaper in the great valley than in the Sates of the Atlantic and the nations of Western Europe, then, and not till then, will it cease to be the interest of artisans and manufacturers to prefer a location in Western towns and cities. This time will probably be about the period when the Mississippi shall flow towards its head.

The chief points for the exchange of the varied productions of industry in our Western valley will necessarily give employment to a great population. Indeed, the locations of our future great cities have been made with reference to their commercial capabilities. Commerce has laid the foundation on which manufactures have been, to a great extent, instrumental in rearing the superstructure. Together, these departments of labor are destined to build up in our fertile valley the greatest cities of the world. J. W. S.

NUMBER V.—1857.

In the rapidly developing greatness of North America, it is interesting to look to the future and speculate on the most probable points of centralization of its commercial and social power. I leave out the political element, because, in the long run, it will not be very potential, and will wait upon industrial developments. I also omit Mexico, so poor, and so disconnected in her relations to the great body of the continent.

Including with our nation, as forming an important part of its commercial community, the Canadas and contiguous provinces, the center of population, white and black, is a little west of Pittsburgh. The movement of this center is north of west, about in the direction of Chicago. The center of productive power cannot be ascertained, with any degree of precision. We know it must be a considerable distance east, and north of the center of population. That center, too, is on the grand march westward. Both, in their regular progress, will reach Lake Michigan. The center of industrial power will touch Lake Erie, and possibly, but not probably, the center of population may move so far northward as to reach Lake Erie also. Their tendency will be to come together; but a considerable time will be required to bring them into near proximity. Will the movement of these centers be arrested before they reach Lake Michigan? I think no one expects to stop eastward of that lake; few will claim that it will go far beyond it. Is it not, then, as certain as anything in the future can be, that the central power of the continent will move to, and become permanent on, the border of the great lakes? Around these pure waters will gather the densest population, and on their borders will grow up the best towns and cities. As the centers of population and wealth approach and pass Cleveland, that city should swell to large size. Toledo will be still nearer the lines of their movement, and should be more favorably affected by them, as the aggregate power of the continent will by that time be greatly increased. As these lines move westward towards Chicago, the influence of their position will be divided between that city and Toledo, distributing benefits according to the degree of proximity.

If we had no foreign commerce, and all other circumstances were equal, the greatest cities would grow up along the line of the central industrial power, in its westward progress, each new city becoming greater than its predecessor, by the amount of power accumulated on the continent, for concentration from point to point of its progress. But as there are points, from one resting-place to another, possessing greatly superior advantages for commerce over all others, and near enough the center line of industrial power to appropriate the commerce which it offers, to these points we must look for our future great cities. To become chief of these, there must be united in them the best facilities for transport, by water and by land. It is too plain to need proof that these positions are occupied by Cleveland, Toledo, and Chicago.

But we have a foreign commerce beyond the continent of North America, by means of the Atlantic Ocean, bearing the proportion, we will allow, of one to twenty of the domestic commerce within the continent. This proportion will seem small to persons who have not directed particular attention to the subject. It is, nevertheless, within the truth. The proof of this is difficult, only because we cannot get the figures that represent the numberless exchanges of equivalents among each other in a community such as ours.

If we suppose ten of the twenty-nine millions of our North American community to earn, on an average, $1 25 per day, 312 days in the year, it will make an aggregate of nearly four thousand millions of dollars. If we divide the yearly profits of industry equally between capital and labor, the proportion of labor would be but $1 25 per day, for five millions of the twenty-nine millions. The average earnings of the twenty-nine millions, men, women, and children, to produce two thousand millions yearly, would be 22 cents a day, for 312 working days. This is rather under than over the true amount; for it would furnish less than $70 each for yearly support, without allowing anything for accumulation.

Of the four thousand millions of yearly production, we cannot suppose that more than one thousand millions is consumed by the producers, without being made the subject of exchange. This will leave three thousand millions as the subjects of commerce, internal and external. Of this, all must be set down for internal commerce, inasmuch as most of that which enters the channel of external commerce first passes through several hands betwen the producer and exporter. Foreign commerce represents but one transaction. The export is sold, and the import is bought with the means the export furnishes. Not so with domestic commerce. Most of the products which are its subjects are bought and sold many times, between the producer and ultimate consumer. Let us state a case:

I purchase a pair of boots from a boot dealer in Toledo. He has purchased them from a wholesale dealer in New York, who has bought them of the manufacturer in Newark. The manufacturer has bought the chief material of a leather dealer in New York, who has made the purchases which fill his large establishment from small dealers in hides. These have received their supply from butchers. The butchers have bought of the drovers, and the drovers of the farmers. If the boots purchased are of French manufacture, they have been the subject of one transaction represented in foreign trade, to-wit: their purchase in Paris by the American importer; whereas, they are the subject of several transactions in our domestic trade. The importer sells them to the jobber in New York; the jobber sells them to the Toledo dealer, who sells them to me.

It can scarcely admit of a doubt that the domestic commerce of North America bears a proportion as large as twenty to one of its foreign commerce. Has internal commerce a tendency to concentrate in few points like foreign commerce? Is its tendency to concentration less than that of foreign commerce? No difference, in this respect, can be perceived. All commerce develops that law of its nature to the extent of its means. Foreign commerce concentrates chiefly at those ports where it meets the greatest internal commerce. The domestic commerce being the great body draws to it the smaller body of foreign commerce. New York, by her canals, her railroads, and her superior position for coastwise navigation, has drawn to herself most of our foreign commerce, because she has become the most convenient point for the concentration of our domestic trade. It is absurd to suppose she can always, or even for half a century, remain the *best* point for the concentration of domestic trade; and, as the foreign commerce will every year bear a less and less proportion to the domestic commerce, it can hardly be doubted that before the end of one century from this time the great center of commerce of all kinds, for North America, will be on a lake harbor. Supposing the center of population (now west of Pittsburgh) shall average a yearly movement westward, for the next fifty years, of twenty miles; this would carry it one thousand miles northwestward from Pittsburgh, and some five hundred or more miles beyond the central point of the natural resources of the country. It would pass Cleveland in five years, and Toledo in eleven years, reaching Chicago, or some point south of it, in less than twenty-five years. The geographical center of industrial power is probably now in Northeastern Pennsylvania, having but recently left the city of New York, where it partially now for a time remains. This center will move at a somewhat slower rate than the center of population. Supposing its movement to be fifteen miles a year, it will reach Cleveland in twenty years, Toledo in twenty-seven years, and Chicago in forty-five years. If ten years be the measure of the annual movement northwestward of the industrial central point of the continent, Cleveland would be reached in thirty years, Toledo in forty, and Chicago in sixty-three years. It is well known that the rate at which the center of population in the United States is now moving westward is over fifteen miles a year, and that it is moving with an accelerated speed. It is obvious that the center of population and the center of industrial power, now widely separated, by the nature of the country between New York and Cleveland, by the superiority in productive power of the old Northern and Middle States over the new States of the Northwest, and still more by the inferiority of industrial power of the plantation States, compared with the region lying north of them, will have a constant tendency to approximate, but can never become identical so long as the inferior African race forms a large portion of the population of the great Southern section of our Union. The constant tendency of the center of industrial power will be northward, as well as westward. This will be determined by the superiority of natural resources of the Northwest over the Southwestern section, by the use of a far greater proportion of machine labor, in substitution for muscular labor, in the northern region, and also by the superior muscular and mental power of the inhabitants of the colder climate. To these might be added the immense advantage of a vastly greater accumulated industrial power in every branch of industry, and the tendency of the superabundant capital of the Old World to flow into the free States and the country north of them.

In the view of the subject which has been taken here, it will be seen that the trade with the British Provinces north of us has been considered a portion of our domestic trade, and that Mexico and California have been left out of our calculation. These may be allowed to balance each other. But, together or apart, they will not be of sufficient importance to our continental commerce to vary materially the results of its future for the next fifty years, as developed in this paper.

At the present rate of increase, the United States and the Canadas fifty years from this time, will contain over one hundred and twenty millions of people. If we suppose it to be one hundred and five millions, and that these shall be distributed so that the Pacific States shall have ten millions, and the Atlantic border twenty-five millions, there will be left for the great interior plain seventy millions. These seventy millions will have twenty times as much commercial intercourse with each other as with the world outside. It is obvious, then, that there must be built up in their midst the great city of the continent; and not only so, but that they will sustain several cities greater than those which can be sustained on the ocean border.

This is the era of great cities. London has nearly trebled in numbers and business since the commencement of the current century. The augmentation of her population in that time has been a million and a half. This increase is equal to the whole population of New York and Philadelphia, and yet it is probable that New York will be as populous as London in about fifty years. A liberal, but not improbable, estimate of the period of duplication of the numbers of these great cities would be, for London thirty years, and for New York fifteen years. At this rate, London will have four millions and seven hundred thousand, and New York three millions four hundred thousand, at the end of thirty years. At the end of the third duplication of New York—that is, in forty-five years—she will have become more populous than London, and number nearly seven millions. This is beyond belief, but it shows the probability of New York overtaking London in about fifty years.

A similar comparison of New York and the leading interior city—Chicago—will show a like result in favor of Chicago. The census returns show the average period of duplication to be fifteen years for New York, and less than four years for Chicago. Suppose that of New York for the future should be sixteen years, and that of Chicago eight years, and that New York now has, with her suburbs, nine hundred thousand, and Chicago one hundred thousand people. In three duplications, New York would contain six millions two hundred thousand, and Chicago in six duplications, occuping the same length of time, would have six millions four hundred thousand. It is not asserted, as probable, that either city will be swelled to such an extraordinary size in forty-eight years—if ever; but it is more than probable that the leading interior city will be greater than New York fifty years from this time.

A few words as to the estimation in which such anticipations are held. The general mind is faithless of what goes much beyond its own experience. It refuses to receive, or it receives with distrust, conclusions, however strongly sustained by facts and fair deductions, which go much beyond its ordinary range of thought. It is especially skeptical and intolerant toward the avowal of opinions, however well founded, which are sanguine of great future changes. It does not comprehend them, and therefore refuses to believe; but it sometimes goes further, and, without examination, scornfully rejects. To seek for the truth is the proper object of those who, from the past and present, undertake to say what will be in the future, and, when the truth is found, to express it with as little reference to what will be thought of it as if putting forth the solution of a mathematical problem.

If we were asked whose anticipations of what has been done to advance civilization, for the past fifty years, have come nearest the truth—those of the sanguine and hopeful, or those of the cautious and fearful—must it not be answered that no one of the former class had been sanguine and hopeful enough to anticipate the full measure of human progress since the opening of the present century? May it not be the most sanguine and hopeful only, who, in anticipation, can attain a due estimation of the measure of future change and improvement in the grand march of society and civilization westward over our continent?

J. W. S.

We have given Mr. Scott the benefit of a full hearing, in order to enable the reader the better to see the justness of the arguments and the truth of the positions in the discussion bearing upon the subject of the pamphlet and the future development of the internal trade of the continent.

No home question of the American people, touching their continental growth and commerce, is so great as this one upon the internal and westward growth of material power. It is the great source of industrial vitality and civil progress.

In the discussion of the internal trade of the continent and the Western movement of the center of population and of the industrial power of North America, Mr. Scott has gone elaborately into the questions, yet he has lived to see some errors in his own arguments; and against them I caution the reader, and point to what I conceive to be the truth in commercial experience and in fact.

The great error of most men who undertake to solve the problems of mankind in the different phases of their career comes first from a failure to draw the correct lessons from history; and second, on account of being too much guided by existing conditions, and not looking beyond to what must be the inevitable unfoldment and growth of their industry from the fixed principles of nature. This was Mr. Scott's error. His reasonings to prove that Toledo would be the great inland center of commerce, and that Chicago and Cleveland would be her handmaids, were founded purely upon the existing condition of things at the time he wrote, while beyond that condition the fixed principles of nature told of a different growth and a different distribution of the commerce of the continent.

Mr. Scott wrote when his vision was circumscribed by the deadening influence of slavery over more than one-half of the States, and when Indian reservations blockaded the way to fertile lands in the West and Southwest. He saw the free States of the North, with their population preponderating in great numbers over the population of the slave States of the South. He saw from those populous States thousands of hardy sons and daughters going forth to the Northwest in search of homes when the way was blockaked to the Southwest, and thus conceived that the life-currents of the nation were destined to localize themselves in that region. At a later day thousands from all parts of the South were seen fleeing from the terrors of the rebellion to the Northwest; an activity and a growth was seen there that has never been equaled on the continent, and short-sighted observers have imagined that all that unparalleled tendency of the people thither, and that extraordinary growth in population and material power, was in conformity to a fixed law of national growth. Not so. These incidental, yet local causes, positively compelled the tide of American progress and power to the lakes and the Northwest. Slavery and Indian titles alone compelled the flank movement of the central column Northward in the civil conquest of the continent. So, too, did the late unhappy war drive the population, the industry, and the wealth, to the Northwest; but, with the extinction of slavery and Indian titles, the continent is left alike all over, and, founded upon the material resources of the country, trade and industry will be guided by the normal action of society and the law of supply and demand, and thus change all the workings of commerce founded alone upon temporary conditions. For each slave set free is added $1,000,000 to the nation's wealth, and for each Indian title extinguished will be added a great community of industrious and intelligent people, who, "yielding to irresistible attraction, will seek a new life in becoming a part of the great whole."

But let us look beyond Mr. Scott's reasoning, and set right those whom he has misguided. Two theories of internal commerce have been written into notice by American writers: one is the Lake theory, and the other is the River theory. The Lake theory has been before the people much the longest time, and has been the subject of a greater number of writers than has the River theory. The Lake theory now is that Chicago is to be the commercial center for the trade of the Mississippi Valley, and that the produce will go there, and from thence over the lakes to New York and foreign markets. The River theory is that the commerce of the Mississippi Valley will follow the rivers to the Gulf, and from thence to the markets of the world. Mr. Scott advocated the Lake theory, first making Toledo the commercial center, but at a later day pointed to Chicago as the favored place. The River theory, as yet, has received but little attention in public print or public enterprise. Although both of these theories are entitled to great consideration by the American people, yet it seems to be but an easy matter to determine which is to be the dominant one. For the Lake theory to prevail, New York must control the commerce of the Valley States and the farther West. This is an utter impossibility. She neither can control it by means of conveyance via the lakes nor by the Gulf. The development of the Valley States and the farther West will break her hold upon this people in spite of her wealth.

It is the commerce going to and from nations that builds great cities on the seaboard, and that, too, when the people of the interior are only a producing people. On the other hand, when a nation has a valuable interior, rich soils, heavy forests, valuable metals, and good water-powers, its people are sure to become a consuming people, and therefore a populous people, and, with the dense population, in the interior, the great cities grow in the interior, and the seaboard cities become scarcely more than shipping ports. France and England give the strongest evidence of this truth. London and Paris are their interior cities, while Liverpool and Brest are their shipping ports. Such will be the result in America. But a few more years and that difference of wealth will not exist between the seaboard cities and those of the West that now does, and, therefore, they cannot exercise that arbitrary commercial control over the trade of the West that they now do.

The rapid approach to the time when our inland cities will equal, and even surpass, the Atlantic cities may be seen in the following figures. Taking the four cities of the seaboard and the four of the interior, they stand thus :

Seaboard Cities.	1860.	Inland Cities.	1860.
Boston	177,840	Cincinnati	161,044
New York	805,651	Chicago	109,260
Philadelphia	565,529	St. Louis	160,773
Baltimore	212,418	New Orleans	168,675
	1,761,438		599,752

Seaboard cities over Western cities.................................. 1,161,686

Seaboard Cities.	1868.	Inland Cities.	1868.
Boston	278,000	Cincinnati	250,000
New York	885,000	Chicago	252,000
Philadelphia	725,000	St. Louis	265,000
Baltimore	230,000	New Orleans	200,000
	2,118,000		967,000

Seaboard cities over Western cities................................... 1,151,000

The figures show a material gain by the inland cities over those of the seaboard, in spite of the ravages of the war upon St. Louis and New Orleans; besides, in the West we have a greater area of country inviting alike all over to the emigrant, which causes a greater diffusion of our Western people than upon the seaboard part of our continent. But give us ten years of peaceful growth, and the West will double in population and wealth. In 1860 St. Louis was the seventh city of the country. She is now the fourth, and will soon be the third.

In another ten years St. Louis will have more railroads running to her than Chicago has. Startling as this statement may be to those who have been for a long time hearing that Chicago was the greatest railroad city in the country, the statement is nevertheless true. Any one who is acquainted with the railroad system of St. Louis, and can comprehend what ten years will bring forth, can see at once the truth of the statement.

In addition to St. Louis becoming the great railroad center, she will command both the Omaha and Kansas Pacific Railroads, for she is more than 100 miles nearer Omaha than Chicago. Besides the road via New Mexico and Arizona to the Pacific ocean must be, on account of climate, the superior road. St. Louis will also have the advantage of the Galveston road and the Mississippi river, which will give her the advantage of the Southern and tropical trade. Thus everywhere are to be seen the unmistakable evidence of the future supremacy of St. Louis and her destiny to become the commercial center of the Mississippi Valley.

On our Western seaboard we have San Francisco, with a population of 125,000, besides many other rapidly growing cities in the interior.

The population of the West will also be more dense than that of the East; also, the workshops and wealth will be greater. Hence the inevitable triumph of the River theory of commerce over the Lake theory. The inhabitants along the rivers will grow the crops, work the metals and the timbers, while the rivers and the railroads bear away over the country and to the Gulf the product of their industry. With cheaper freights and greater advantages, resulting from greater proximity to the produce, the River theory must prevail, and the interest of Chicago, St. Louis, and New Orleans be one in the united industrial and commercial movements of the West.

He who reasons for the results of the future must take for the basis of his arguments the facts as they exist in nature as well as in man, and combine them in proper relations, and then he becomes a prophet among his people. Man's success everywhere comes from his working in harmony with nature's laws. Then, in conformity to these overruling conditions, the commerce of the Mississippi Valley must follow the flow of the rivers, and the wealth of the people must come from the soils, the minerals, and the forests. In response to all these truths, the River theory of the commerce of the West must be dominant over the Lake theory. In support of this position, the following facts are offered as still greater evidence of its truth:

The States lying upon the banks of the Ohio and Mississippi rivers, fourteen in number, had, by the census of 1860, a population of 16,909,494, or more than half the whole population of the United States; and these two rivers have a coast line of 36,098 miles, while the coast of the Atlantic is 2,163 miles, and the Gulf of Mexico 1,764 miles, and of the Pacific 1,348 miles, on an outer line, or 21,354 miles including bays and indentations.

That these rivers drain an area of 1,785,267 square miles, more than half of the whole 3,001,002 square miles in the United States; and these fourteen States, in 1860, contained 94,402,869 of the 163,110,720 improved acres, and 126,703,893 of the 244,101,818 unimproved acres of the whole United States; and the valuation of property in these fourteen States shows $8,467,511,274 of the whole valuation of the United States, $16,077,358,715; showing very conclusively that these fourteen States pay more than half the taxes, work more than half of the improved land, have the majority of the population, and also the majority of the land to develop, of the whole United States.

By the census of 1860, the whole product of the United States was valued at $1,900,000,000, while the foreign exports of domestic produce were only $373,189,274, or less than one-fifth of the whole product, leaving four-fifths for exchange in domestic commerce between the States.

The proportion of the whole product afforded by these fourteen States we speak for, may be judged by these returns of their produce, gathered from the census of 1860, and compared with the whole United States, as follows:

	The Fourteen States.	The whole United States.
Corn	632,453,375 bushels.	838,792,740 bushels.
Wheat	126,920,730 "	173,104,924 "
Oats	103,995,461 "	172,643,185 "
Tobacco	345,400,759 pounds.	434,209,461 pounds.
Sugar	222,636,000 "	230,982,000 "
Cotton	1,079,799,600 "	2,154,820,800 "
Wool	31,277,839 "	60,264,913 "
Hay	9,297,743 tons.	19,083,896 tons.
Butter	239,601,405 pounds.	459,681,372 pounds.
Hemp	69,470 tons.	74,493 tons.
Hogs	22,225,766	31,512,867
Bituminous coal	3,247,264,425 bushels.	3,621,923,165 bushels
Horses and asses	4,804,634	7,400,322
Cattle	12,517,392	25,616,019
Sheep	11,973,315	22,471,275

Showing for the river States a great preponderance in the products of the whole country.

The total tonnage owned in the United States is returned in the census of 1860 as 5,353,868 tons, and the portion belonging to the fourteen States as 996,266 tons; but it is estimated, by competent parties, that the steamers on the Ohio and Mississippi have carried 7,905,216 tons during the year 1866, evincing the activity in domestic commerce of these river States, and this commerce but yet in its infancy — for it is developing daily, and demonstrating that from these States has and must come the food supply for the whole nation and for export; and that they must also supply the gold and silver States which are developing so largely and quickly upon the tributaries of their rivers.

These figures cannot be regarded otherwise than in favor of the River theory, and the consequent development of St. Louis as the commercial center of the Mississippi Valley.

In further proof of St. Louis becoming the commercial center of the Mississippi Valley, the following evidence given by Professor Waterhouse, of this city, in one of his valuable articles upon the resources of Missouri, is submitted:

ST. LOUIS THE COMMERCIAL CENTRE OF NORTH AMERICA.

St. Louis is ordained by the decrees of physical nature to become the great inland metropolis of this continent. It cannot escape the magnificence of its destiny. Greatness is the necessity of its position. New York may be the head but St. Louis will be the heart of America. The stream of traffic which must flow through this mart will enrich it with alluvial deposits of gold. Its central location and facilities of communication unmistakably indicate the leading part which this city will take in the exchange and distribution of the products of the Mississippi Valley. St. Louis is situated upon the west bank of the Mississippi, at an altitude of 400 feet above the level of the sea. It is far above the highest floods that ever swell the Father of Waters. Its latitude is 38 deg. 37 min. 28 sec. north, and its longitude 90 deg. 15 min. 16 sec. west. It is 20 miles below the mouth of the Missouri, and 200 above the confluence of the Ohio.

				Miles.
Distance by river from St. Louis to	Keokuk	200
"	"	"	Burlington	260
"	"	"	Rock Island	350
"	"	"	Dubuque	470
"	"	"	St. Paul	800
"	"	"	Cairo	200
"	"	"	Memphis	440
"	"	"	Vicksburg	830
"	"	"	New Orleans	1,240
"	"	"	Louisvill	580
"	"	"	Cincinnati	720
"	"	"	Pittsburg	1,200
"	"	"	Leavenworth	500
"	"	"	Omaha	800
"	"	"	Sioux City	100
"	"	"	Fort Benton	3,100
Distance by rail from St. Louis to	Indianapolis	200
"	"	"	Chicago	280
"	"	"	Cincinnati	340
"	"	"	Cleveland	470
"	"	"	Pittsburg	650
"	"	"	Buffalo	650
"	"	"	New York	1,000
"	"	"	Lawrence	320
"	"	"	Denver	880
"	"	"	Salt Lake	1,300
"	"	"	Virginia City	1,900
"	"	"	San Francisco	2,300

St. Louis very nearly bisects the *direct* distance of 1,400 miles between Superior City and the Balize. It is the geographical center of a valley which embraces 1,200,000 square miles. In its course of 3,200 miles, the Mississippi borders upon Missouri 470 miles. Of the 3,000 miles of the Missouri, 500 lie within the limits of our own State. St. Louis is mistress of more than 16,500 miles of river navigation.

This metropolis, though in the infancy of its greatness, is already a large city. Its length is about eight miles, and its width three. Suburban residences, the outposts of the grand advance, are now stationed six or seven miles from the river. The present population of St. Louis is 204,300. In 1865, the real and personal property of the city was assessed at $100,000,000, and in 1866 at $126,877,000.

St. Louis is a well-built city, but its architecture is rather substantial than showy. The wide, well-paved streets, the spacious levee, and commodious warehouses; the mills, machine-shops, and manufactories; the fine hotels, churches, and public buildings; the universities, charitable institutions, public schools, and libraries, constitute an array of excellences and attractions of which any city may justly be proud. The Lindell and Southern Hotels are two of the largest and most magnificent structures which the world has ever dedicated to public hospitality. The Lindell is itself a village.*

The appearance of St. Louis from the eastern bank of the Mississippi is impressive. At East St. Louis, the eye sometimes commands a view of 100 steamboats lying at our levee. Notwithstanding the departure of more than 40 boats for Montana, there are at this date 70 steamers in the port of St. Louis. A mile and a half of steamboats is a spectacle which naturally inspires large views of commercial greatness. The sight of our levee, thronged with busy merchants and covered with the commodities of every clime, from the peltries of the Rocky Mountains to the teas of China, does not tend to lessen the magnitude of the impression.

The growth of St. Louis, though greatly retarded by social institutions, has been rapid. The population of the city was, in

1769	891	1837	12,040
1795	925	1840	16,469
1810	1,400	1844	34,140
1820	4,928	1850	74,439
1828	5,000	1852	94,000
1830	5,852	1856	125,200
1833	6,397	1859	185,587
1835	8,316	1866	204,327

*On the 30th of March, 1867, this superb edifice was burned to the ground.

In 1866, 1,400 buildings, worth $3,500,000, were erected in St. Louis. The total number of structures in the city is now about 20,000, and their approximate value is $50,000,000.*

At the present rate of decennial increase, St. Louis, in 1900, would contain more than 1,000,000 inhabitants. This number certainly seems to exceed the present probability of realization, but the future growth of St. Louis, vitalized by the mightiest forces of a free civilization, and quickened by the exchanges of a continental commerce, ought to surpass the rapidity of its past development.

The real estate in St. Louis was, in

1859 assessed at	$69,846,845	1863 assessed at	$49,409,030
1860 " "	73,765,670	1864 " "	53,205,820
1861 " "	57,537,415	1865 " "	73,960,700
1862 " "	40,240,450	1866 " "	81,961,010

In 1866, the valuation of the real and personal property in St. Louis on which the State and military taxes were levied was $126,877,000.

The amount of duties collected at the St. Louis Custom House was, in

1861	$30,183 96	1864	$76,448 43
1862	20,404 70	1865	586,407 47
1863	36,622 09	1866	785,052 30

The amount of imposts paid at the port of Chicago during the fiscal year ending December 31, 1866, was $509,643 89 in coin.

The duties collected during the same period at this port amounted to $60,176 45 in currency, and $780,706 97 in gold.

Only about one-fifth of the customs levied on goods imported into St. Louis are collected at this point. St. Louis is only a port of delivery. The imposts upon our foreign merchandise are chiefly paid at the ports of entry.

The present system of foreign importation is unfavorable to the commercial interests of St. Louis. This city should be made a port of entry. The goods of St. Louis importers are now subjected to great delay and expense at New Orleans. The municipal authorities do not permit the merchandise to lie on the landing more than five days. If the requisite papers are not made out within that time, the goods are sent to bonded warehouses. This contingency not unfrequently occurs. The press of business or official slowness often delays the issue of the Custom House pass beyond the specified time, and then the

---

* A report recently made under municipal authority, shows that at the date of the present publication, November, 1868, more than 2,000 buildings—almost all of them built of brick, and many of them faced with stone—are either now in process of erection or just finished.

Western importer is subject to the serious expense which the drayage to the warehouse, loss of time, and frequent damage to the goods, involve. The gravity of this embarrassment forces many of our merchants to pay the duties at New Orleans. This course saves delay and expense. The revenue laws recognize no distinction between the actual payment of duties and the transportation bond. But practically there is an important difference. In case the impost is paid at New Orleans, the goods are almost always forwarded within five days; but when the merchandise is shipped under a transportation bond, the detention is very frequently ten days, and sometimes a month. In the former instance, any package can be forwarded as soon as the duty is paid; but, in the latter case, the imports cannot be dispatched to their destination till the entire shipment has passed the inspection of the Custom House. In consequence of these unjust discriminations against St. Louis, many of our largest importers, notwithstanding the inconvenience of keeping gold on deposit in New Orleans, prefer to pay the duties on their foreign goods at the port of entry.

An excessive and unnecessary delay at the New Orleans Custom House recently subjected one of our merchants to a loss of $8 a ton on a shipment of iron.

Last season, another of our importers ordered a large stock of Christmas goods. The articles reached New Orleans in season, but were detained there till after the holidays. They must now be kept, with loss and deterioration, for another year; and, before next Christmas, they may become comparatively worthless by changes of mode and new directions of public taste.

These examples illustrate the importance of time in commercial transactions.

The Government could easily obviate all the difficulties which our importers now experience by making St. Louis a port of entry. The commercial embarrassments of the present system need immediate removal. In the event of the proposed change, frauds upon the Government could be prevented by reshipping the goods at New Orleans under the eye of the Custom House authorities, keeping them during the voyage under lock and key, and, if necessary, subjecting them on the passage to the surveillance of a Revenue officer. During the rebellion, the shipments of merchandise to Southern ports were placed under similar supervision. The satisfactory operation of this system, amid all the liabilities to abuse which exist in times of civil turbulence, warrants the conviction that the proposed plan would, in a period of peace, prove eminently successful.

If Congress respects commercial rights, St. Louis will soon become a port of entry.

From the records of the United States Assessor, it appears that in 1865 the sales of 612 St. Louis firms amounted to $140,688,856. For the same year, the imports of this city reached an aggregate of $235,873,875.

The manufactures of St. Louis constitute an important element in our commercial transactions. In 1860, the capital invested in manufactures was $9,205,205, and the value of the product was $21,772,323. In 1866, the mills of this city made 820,000 barrels of flour.

In 1865, our receipts of grain, including flour, were...17,637,250 bushels.
" 1866,   "   "   "   ...20,855,280   "
" 1865, exports . "   "   ...13,427,000   "
" 1866,   "   "   "   ...18,680,500   "

St. Louis, though the eighth city in the United States in population, ranks as seventh in the importance of its manufactures. Missouri might profitably imitate the activity of its metropolis.

The extent of our social and commercial intercourse with the rest of the world may be inferred from the postal statistics of this department. In 1865, the number of letters which passed through the St. Louis Post Office for distribution, mail or delivery, was about 11,000,000. In 1866, the total sum of postage collected, including the sale of stamps, was more than $195,000; and the amount of money orders paid was $145,000. In postal importance, St Louis is the fifth city of the Union.

The earnings of our railroads indirectly exhibit the magnitude of our trade. For the fiscal year of 1865 the total receipts of the Iron Mountain were $424,700; North Missouri, $1,013,000; Missouri Pacific and Southwest Branch, $1,939,000; Hannibal and St. Joseph, $2,000,000. In 1866, the earnings of the Missouri Pacific were $2,670,000. The returns of the Union Pacific for November, 1866, were $77,869. The Directors estimate their monthly receipts for 1867 at $100,000.

In 1865, the total number of passengers, by river or rail, who made St. Louis their destination or a point of transit, amounted to 1,180,000; and in 1866, 1,250,000.

In 1866, the number of houses and firms doing business in St. Louis was 5,500, and the number of commercial licenses issued during the same year was 4,800.

The tonnage owned and enrolled in the district of St. Louis in 1865 was 97,000 tons. On the first of January, 1867, the amount of our steam tonnage, exclusive of a large number of barges and canal boats which made occasional trips, was 106,600 tons, with a carrying capacity of 186,000 tons, and a value of $10,376,000.

Our commerce is aided by ample banking facilities. There are in St. Louis, in addition to 20 private banks, 38 insurance companies, 31 incorporated banking institutions, with an actual capital of $15,000,000. The character of our banks stands deservedly high in the financial world. The development of the territories is bringing large deposits to our banks, creating new demands for capital, and extending the channels of circulation.

Our trade with the mountains is large and rapidly increasing. In 1865, 20 boats set out from this port for Fort Benton—which is more than 3,000 miles from St. Louis—with a total freight of 6,000,000 pounds.

In 1866, 50 boats sailed for Fort Benton, with an aggregate tonnage of 10,284 tons. In three instances the cost of assorted goods was as follows:

13 tons of merchandise	$12,000
35 " "	40,000
40 " "	65,000
Mean cost per ton	1,300

The agent who furnishes these facts feels authorized by his experience in the trade of the Upper Missouri to appraise a ton of Montana merchandise at $1,000.

The following table is an approximate estimate, based upon the preceding data, of our commerce with Montana for the year 1866:

Number of boats	50
" " passengers	2,500
Pounds of freight	13,000,000
Value of merchandise	$6,500,000

The trade across the Plains is of still greater magnitude. The overland freight from Atchison alone has increased from 3,000,000 pounds in 1861 to 21,500,000 in 1865.

The Overland Dispatch Company have courteously furnished me with estimates, founded upon their own transactions, of our total commerce with the Territories in 1865. These figures do not include the Fort Benton trade.

Number of passengers East and West by overland coaches	4,800
" " " " by trains and private conveyances	50,000
Number of wagons	8,000
" " cattle and mules	100,000
Pounds of freight to Plattsmouth	3,000,000
" " Leavenworth City	6,000,000
" " Santa Fe	8,000,000
" " St Joseph	10,000,000
" " Nebraska City	15,000,000
" " Atchison	25,000,000
Government freight	50,000,000
Total number of pounds	117,000,000
Amount of treasure carried by express	$3,000,000
" " " by private conveyance	30,000,000

The Overland Express charge 3 per cent. for the transportation of bullion. This high commission and the hostility of the Indian tribes induced many miners to send their gold East by the way of San Francisco to Panama.

In 1866, the total assay of bullion in the United States was $81,389,540. Of this aggregate, $73,032,800 came from the Pacific and Rocky Mountain mines. Upon the usual estimate that 25 per cent. of the gold and silver escapes assay, the entire product of the country in 1866 was $100,000,000. The increase of population in the gold regions, the richness of recent discoveries, and greater activity in mining operations indicate a still larger aggregate in 1867.

In 1866, the Westward traffic of Leavenworth amounted to $50,000,000. This aggregate includes the Santa Fe trade, whose value last year was about $35,000,000. The Western trade of Nebraska City was, in

1863	16,800,000 pounds.
1864	23,000,000 "
1865	44,000,000 "
1866	30,000,000 "

The freightage from this point across the Plains required, in 1865, 11,739 men, 10,311 wagons, 10,123 mules, and 76,596 oxen.

So great is the length of the overland routes that the trains are able to make but two through trips a year.

The Union Pacific railroad already extends to Fort Harker. This materially shortens the extent of overland freightage.*

Distance from St. Louis to Fort Harker	508 miles.
" " Fort Harker to Denver	372 "
" " " " Salt Lake City	860 "
" " " " Virginia City	1432 "

The length of these lines of transportation, the slowness of our present means of communication, and the magnitude of our territorial population and trade, forcibly illustrate the necessity of a Pacific railroad.

The foregoing summaries exhibit the commerce of the Mississippi Valley with the mountains. But while St. Louis does not monopolize the trade of the gold regions, it yet sends to the

---

* The Union Pacific, Eastern Division, now extends to Sheridan, 688 miles west of St. Louis. The distance from Sheridan to Denver is 175 miles, and from Denver to Cheyenne—where the Union Pacific forms a junction with the Northern line—112 miles.
The Northern Pacific is now completed 850 miles west of Omaha. The Central Pacific now runs eastward from San Francisco more than 600 miles. The 400 miles which remain to be built will probably be finished by the fourth of July, 1869—more than six years before the time prescribed by law for the completion of the road. Then an unbroken line of railway of 3,300 miles long will stretch from New York to San Francisco. This gigantic work, prosecuted during the most formidable rebellion of modern times, and finished amid the derangements of national finance incident to civil convulsions, must ever be regarded as an extraordinary triumph of American energy.

territories by far the largest portion of their supplies. Even in cases where merchandise has been procured at intermediate points, it is probable that the goods were originally purchased at St. Louis.

During the rebellion, the commercial transactions of Cincinnati and Chicago doubtless exceeded those of St. Louis. The very events which prostrated our trade stimulated theirs into an unnatural activity. Their sales were enlarged by the traffic which was wont to seek this market. Our loss was their gain.

The Southern trade of St. Louis was utterly destroyed by the blockade of the Mississippi. The disruption by civil commotions of our commercial intercourse with the interior of Missouri was nearly complete. The trade of the Northern States, bordering upon the Mississippi, was still unobstructed. But the merchants of St. Louis could not afford to buy commodities which they were unable to sell, and country dealers would not purchase their goods where they could not dispose of their produce. Thus St. Louis, with every market wholly closed or greatly restricted, was smitten with a commercial paralysis. The prostration of business was general and disastrous. No comparison of claims can be just which ignores the circumstances that, during the rebellion, retarded the commercial growth of St. Louis, yet fostered that of rival cities.

Nothing more clearly demonstrates the geographical superiority of St. Louis than the action of the Government during the war. Notwithstanding the strenuous competition of other cities, our facilities for distribution and a due regard for its own interests compelled the Government to make St. Louis the Western base of supplies and transportation. During the rebellion, the transactions of the Government at this point were very large. General Parsons, Chief of Transportation in the Mississippi Valley, submits the following as an approximate summary of the operations in his department from 1860 to 1865:

AMOUNT OF TRANSPORTATION.

Cannons and caissons	800
Wagons	13,000
Cattle	80,000
Horses and mules	250,000
Troops	1,000,000
Pounds of military stores	1,950,000,000

General Parsons thinks that full one-half of the transportation employed by the Government on the Mississippi and its tributaries was furnished by St. Louis.

From September 1, 1861, to December 31, 1865, General Haines, Chief Commissary of this department, expended at St. Louis, for the purchase of subsistence stores, $50,700,000.

During the war, General Myers, Chief Quartermaster of this department, disbursed at this city, for supplies, transportation, and incidental expenses, $180,000,000.

The national exigencies forced the Government to select the best point of distribution. The choice of the Federal authorities is a conclusive proof of the commercial superiority of St. Louis.

The conquest of treason has restored to this mart the use of its natural facilities. Trade is rapidly regaining its old channels. On its errand of exchange it penetrates every State and Territory in the Mississippi Valley, from Alabama and New Mexico to Minnesota and Montana. It navigates every stream that pours its tributary waters into the Mississippi. It visits the islands of the sea, traverses the ocean, and explores foreign lands.

Before the war, almost all the Western trade in coffee and sugar was carried on by way of New Orleans. The interruption of traffic, by the blockade of the Mississippi river, changed the channels of commerce. By the necessities of the country, trade was forced into unnatural courses. New York, by its limitless capital and enterprise, has obtained a brief control over a trade that rightfully belongs to the West. As soon as the country regains its normal condition and commerce resumes its natural flow, the West will inevitably assert its former and legitimate ascendancy in this branch of business. Most of the coffee used in the West is brought from Rio Janeiro. Water carriage is always the cheapest means of transportation. The rail from New York cannot compete with the river from New Orleans. Besides, the Gulf route is the shortest distance between St. Louis and Rio Janeiro. The cost, then, of importing Rio coffee to this point is much less by New Orleans than by New York. An urgent necessity exists for the establishment of lines of steamers between New Orleans and South American ports.

A direct trade with the West Indies and South America would, from our superior facilities of transportation, not only place the control of the grocery business of the Northwest in our hands, but also greatly enlarge our exportations. The West consumes far more coffee proportionately than the East. South America uses large quantities of Western flour. There would then be a steady and growing interchange of commodities between these countries.

Missouri flour is the best in the American market. This is an important advantage in favor of St. Louis. It is a well-ascertained fact that flour made from grain grown in this latitude bears the voyage to South American ports better than any other. The experience of exporters verifies this assertion. Our flour is, then, not only the finest in the United States for home consumption, but also the best for exportation to tropical countries.

St. Louis ought to cultivate more intimate commercial relations with Brazil. Prior to our acquisition of Russian America, the area of this country was 500,000 square miles larger than that of the United States. Its present population is nearly 10,000,000. Of its principal cities,

Para contains	30,000 inhabitants.
Pernambuco	80,000 "
Bahia	130,000 "
Rio Janeiro	400,000 "

The exports of Brazil are coffee, hides, sugar, caoutchouc, rosewood, mahogany, Brazil wood, cinchona, logwood, cotton, rice, sarsaparilla, sassafras, ipecacuanha, cacao, vanilla, cloves, cinnamon, and tamarinds.

In 1856, the value of the commodities imported from Brazil into the United States was—

Brazil wood	$32,000
" nuts	43,000
Rosewood	81,460
Hair	138,240
Sugar	513,450
India rubber	771,320
Raw hides	1,930,220
Coffee	16,091,700

In 1857, this country imported from Brazil 197,000,000 pounds of coffee, worth $17,980,000. In the same year Brazil exported to foreign markets 256,000,000 pounds of sugar.

In exchange for these valuable commodities, Brazil needs lard, pork, hams, flour, pine lumber, agricultural implements, textile fabrics, and other manufactures. These articles are the chief staples of Western growth and production. The Mississippi Valley is able to supply most of the commercial wants of Brazil. St. Louis, as the main distributing point of the West, ought to take the lead in this grand system of mercantile exchanges. A vast commerce must soon spring up between the metropolis of this valley and the ports of South America. But at present our exports to Brazil are entirely disproportioned to our ability to meet the commercial wants of that country. In 1854-55, the trade of England with South America was five times as large as that of the United States.

In 1830, the value of our American imports from Brazil was...$20,000,000
"   "   "   " exports   "   " ... 6,000,000

These figures show that this country is not a successful competitor for the rich trade of South America. More energetic rivals are enriching themselves with the opulence of this commerce.

The wants of the United States and Brazil are complementary. Each country needs the productions of the other. The West is the fruitful and main source of those commodities which South America requires. St. Louis, as the chief emporium of the Mississippi Valley, is able, by the vast expansion which it can cause in this tropic trade, to turn the commercial balance in favor of the United States, and itself become the central distributing point of Brazilian staples.

But St. Louis can never realize its splendid possibilities without effort. The trade of the vast domain lying east of the Rocky Mountains and south of the Missouri river is naturally tributary to this mart. St. Louis, by the exercise of forecast and vigor, can easily control the commerce of 1,000,000 square miles. But there is urgent need of exertion. Chicago is an energetic rival. Its lines of railroad pierce every portion of the Northwest. It draws an immense commerce by its network of railways. The meshes which so closely interlace all the adjacent country gather rich treasures from the tides of commerce. Chicago is vigorously extending its lines of road across Iowa to the Missouri river. The completion of these roads will inevitably divert a portion of the Montana trade from this city to Chicago. The energy of an unlineal competitor may usurp the legitimate honors of the imperial heir.

St. Louis cannot afford to continue the masterly inactivity of the old *regime*. A traditional and passive trust in the efficacy of natural advantages will no longer be a safe policy. St. Louis must make exertions equal to its strength and worthy of its opportunities. It must not only form great plans of commercial empire, but must execute them with an energy defiant of failure. It must complete its projected railroads to the mountains, and span the Mississippi at St. Louis with a bridge whose solidity of masonry shall equal the massiveness of Roman architecture, and whose grandeur shall be commensurate with the future greatness of the Mississippi Valley. The structure whose arches will bear the transit of a continental commerce should vie with the great works of all time, and be a monument to distant ages of the triumph of civil engineering and the material glory of the Great Republic.

Since these sentences were written, a company, composed of men of large means and sterling integrity, has been incorporated for the purpose of erecting a bridge across the Mississippi at this point. The executive and financial ability of its members is a guarantee of efficient action and an early accomplishment of this great work. The length of the bridge, together with its approaches, will be about 3,500 feet, and the probable cost $5,000,000. The material of the structure will be steel.

Chas. K. Dickson is president of the company, and James B. Eads, the distinguished inventor, is chief engineer.

The initial steps for the erection of a bridge across the Missouri at St. Charles have already been taken. The work should be pushed forward with untiring energy to its consummation.

The iron, stone, and timber necessary for these structures can be obtained within a few miles of St. Louis, and the greater part of the material can be transported by water. The construction of public works whose cost would be millions of dollars would afford employment to thousands of laborers, and give fresh impulse to the prosperity of St. Louis.

A full and persistent presentation of the superior claims of Carondelet ought to induce the Government to establish a naval station at that point. The supply of labor and *materiel* which a navy yard would require would be another source of wealth to Missouri and its metropolis.

The effect of improvements upon the business of the city may be illustrated by the operations of our city elevator. The elevator cost $450,000, and has a capacity of 1,250,000 bushels. It is able to handle 100,000 bushels a day. It began to receive grain in October, 1865. Before the first of January, 1866, its receipts amounted to 600,000 bushels, 200,000 *of which were brought directly from Chicago.* The total receipts at the elevator in 1866 were 1,376,700 bushels. Grain can now be shipped, by way of St. Louis and New Orleans, to New York and Europe twenty cents a bushel cheaper than it can be carried to the Atlantic by rail.

The facilities which our elevator affords for the movement of cereals have given rise to a new system of transportation. The Mississippi Valley Transportation Company has been organized for the conveyance of grain to New Orleans in barges. Steam tugs of immense strength have been built for the use of the company. They carry no freight. They are simply the motive power. They save delay by taking fuel for the round trip. Landing only at the large cities, they stop barely long enough to attach a loaded barge. By this economy of time and steady movement, they equal the speed of steamboats. The Mohawk made its first trip from St. Louis to New Orleans in six days, with ten barges in tow. The management of the barges is precisely like that of freight cars. The barges are loaded in the absence of the tug. The tug arrives, leaves a train of barges, takes another, and proceeds. The tug itself is always at work. It does not lie idle at the levee while the barges are loading. Its longest stoppage is made for fuel. The power of these boats is enormous. The tugs plying on the Minnesota river sometimes tow 30,000 bushels of wheat apiece. The freight of a single trip would fill 85 railroad cars.

Steamboats are obliged to remain in port two or three days for the shipment of freight. The heavy expense which this delay and the necessity for large crews involve is a grave objection to the old system of transportation. The service of the steam tug requires but few men, and the cost of running is relatively light. The advantages which are claimed for the barge system are exhibited by the following table:

	Tugs and barges.	Steamboats.
Stoppage at intermediate points	2 hours	6 hours
" " terminal	24 "	48 "
Crew	15	50
Tonnage	25,000 tons	1,500 tons
Daily expense	$200	$1,000
Original cost	$75,000	$100,000

In addition to the ordinary precautions against fire, the barges have this unmistakable advantage over steamboats: they can be cut adrift from each other, and the fire restricted to the narrowest limits. The greater safety of barges ought to secure for them lower rates of insurance. The barges are very strongly built, and have water-tight compartments for the movement of grain in bulk. The transportation of grain from Minnesota to New Orleans by water costs no more than the freightage from the same point to Chicago. After the erection of a floating elevator at New Orleans, a boat load of grain from St. Paul will not be handled again till it reaches the Crescent City.

At that port it will be transferred by steam to the vessel which will convey it to New York or Europe. The possible magnitude of this trade may be inferred from the fact that in 1865 Minnesota alone raised 10,000,000 bushels of wheat. Three quarters of this harvest could have been exported if facilities of cheap transportation had offered adequate inducement. In 1866, higher prices — which produced the same practical result as cheaper freightage — led to the exportation of 8,000,000 bushels. Some of this grain belonged to the crop of the preceding year. But this fact does not at all affect the question of carriage.

From the 1st of May to the 25th of December, 1866, the towboats of this city transported 120,000 tons of freight. This new scheme of conveying freight by barges bids fair to revolutionize the whole carrying trade of our Western waters. It will materially lessen the expense of heavy transit, and augment the commerce of the Mississippi river in proportion to the reduction it effects in the cost of transportation. The improvement which facilitates the carriage of our cereals to market, and makes it more profitable for the farmer to sell his grain than to burn it, is a national benefit. This enterprise, which may yet change the channel of cereal transportation, shows what great results a spirit of progressive energy may accomplish.

The mercantile interests of the West imperatively demand the improvement of the Mississippi and its main tributaries. This is a work of such prime and transcendent importance to the commerce of the country that it challenges the co-operation of the Government. A commercial marine which annually transfers tens of millions of passengers, and cargoes whose value is hundreds of millions, ought not to encounter obstructions which human effort can remove. The yearly loss of property from the interruption of communication and wreck of boats reaches a startling aggregate.

For the accomplishment of an undertaking so vital to its municipal interests, St. Louis should exert its mightiest energies. The prize for which competition strives is too splendid to be lost by default. The Queen City of the West should not voluntarily abdicate its commercial sovereignty.

If the emigrant merchants of America and Europe, who recognize in the geographical position of St. Louis the guarantee of mercantile supremacy, will become citizens of this metropolis, they will aid in bringing to a speedier fulfillment the prophecies of its greatness. The current of Western trade must flow through the heart of this valley.

In the march of progress St. Louis will keep equal step with the West. Located at the intersection of the river which traverses zones, and the railway which belts the continent, with divergent roads from this centre to the circumference of the country, St. Louis enjoys commercial advantages which must inevitably make it the greatest inland emporium of America. The movement of our vast harvests and the distribution of the domestic and foreign merchandise required by the myriad thousands who will, in the near future, throng this valley, will develop St. Louis to a size proportioned to the vastness of the commerce it will transact. This metropolis will not only be the centre of Western exchanges, but also, if ever the seat of government is transferred from its present locality, the capital of the nation.

St. Louis, strong with energies of youthful freedom, and active in the larger and more genial labors of peace, will greet the merchants of other States and lands with a friendly welcome, afford them the opportunities of fortune, and honor their services in the achievement of its greatness.

All must agree upon the fact of the wonderful growth of the internal commerce of the country, but difference of opinion may exist as to how that commerce will be distributed so as to build up wealthy and powerful cities and peoples. This is easily determined. In the very nature of things, the tropical climates combine with the temperate to produce wealth and sustenance for man, but never to any success do the temperate climates combine with the frozen climates to produce wealth and commerce. Besides those combinations, the rivers in all lands that have been serviceable to man essentially flow to the tropics. These facts all combine in favor of St. Louis; and surrounded every way by navigable rivers, approached by railways, and in the very midst of the finest coal and iron fields in the world, a destiny of unspeakable greatness is thrust upon her, and also upon the West. The Illinois coal fields are estimated by Prof. H. F. Rodgers to contain 1,227,500,000,000 tons, while the Pennsylvania coal fields contain 316,400,000,000 tons. The Missouri coal fields are estimated by Prof. G. W. Swallow at 109,500,000,000 tons, and yet, owing to the incomplete geological survey of the State, it is thought by competent men that there is still more coal in Missouri. All the coal fields of North America are estimated at 4,000,000,000,000 tons; the coal fields of Great Britain at 190,000,000,000 tons. The Illinois coal fields contain four times as much coal as those of Pennsylvania, nearly one-third as much as all those of North America, and over six times as much as all the coal fields of Great Britain. It is reckoned by Prof. Forrest Shepherd that the best coal fields of Illinois are situated along the Mississippi river, near the southwestern boundary, and adjacent to the Missouri iron fields; that Illinois coal will have to be used in the manufacture of Missouri iron; and that the day is not distant when one vast series of iron foundries and workshops will line the Illinois and Missouri shores of the great river; and thus from Illinois and Missouri will grow, within one hundred to one hundred and fifty miles south of St. Louis, greater shops than the minerals of England have produced. Nothing is more certain than this; for all that nature can do for man she has done in America, and localized it in the Mississippi Valley. Within 100 miles of St. Louis, gold, iron, lead, zinc, copper, tin, silver, platina, nickel, emery, cobalt, coal, lime stone, granite, pipe clay, fire clay, marble, metallic paints, and salt, are found, all of which will repay for working, and most of which are in great abundance. Iron everywhere in civilized life is more valuable than gold. In connection with the consideration of the development of the internal trade of the continent, it is plain to be seen that the interior cities, Chicago, Cincinnati, St. Louis, and New Orleans, are destined to approach, if not rival, Boston, New York, Philadelphia, and Baltimore, in wealth, commerce, population, and material power.

Turn which way we will, at home or abroad, and everything points to the future development and population of the Valley States to immeasurable greatness—the home of more millions of intelligent, industrious, and sovereign people than now live upon the globe.

The agricultural growth of the Northwestern Valley States may be inferred from the following tables deduced from the United States census of 1860:

In 1860 the whole number of acres of improved land in all the States and Territories was 163,261,889. Of this—

Missouri contains	6,246,871
Illinois	13,254,473
Iowa	3,780,253
Wisconsin	3,746,036
Minnesota	554,397
Or a fraction less than one-sixth.	27,582,030

The total value of crops for 1864 is estimated by the Agricultural Bureau of the Department of the Interior to have been $1,564,543,690. Of this sum—

Illinois produced	$214,488,426
Wisconsin	51,938,952
Missouri	52,996,592
Iowa	71,100,481
Minnesota	13,168,123
	$403,692,574

Or more than one-fourth of the value of the entire crops of the country. But these estimates of value are the estimated value of the various products in the States where produced.

Of the value of the live stock, which, on the 1st of January, 1865, was $990,876,128—

Illinois had	$116,588,288
Missouri	44,431,766
Iowa	66,572,496
Wisconsin	36,911,165
Minnesota	8,860,015
Or more than one-fourth.	$273,363,730

A juster standard by which to measure the productiveness of these States would be a comparison of the amount of their respective products, since the value is so largely affected by the distance from market.

The great staples of agriculture are wheat, corn, beef, and pork. Comparing these, we find that the total number of bushels of wheat produced in all the States and Territories in 1864 (except the cotton States, whose production was almost nominal, probably not more than one-sixth of what it was in 1860), was 160,695,823 bushels, of which—

Illinois produced	33,371,173
Missouri	3,281,514
Wisconsin	14,168,317
Iowa	12,649,807
Minnesota	2,634,975
	66,105,786

Or a fraction less than one-half.

The total number of bushels of corn produced was 530,451,403.

Illinois produced	138,356,135
Missouri	36,635,011
Wisconsin	10,087,053
Iowa	55,261,240
Minnesota	4,647,329
	244,986,768

Or nearly one-half.

The whole number of cattle and oxen, January 1, 1865, was 7,072,591.

Illinois had	978,700
Missouri	471,006
Wisconsin	388,760
Iowa	561,338
Minnesota	127,175
	2,526,979

Or more than one-third.

The total number of hogs was 13,070,887.

Illinois had	2,034,231
Missouri	988,857
Wisconsin	340,638
Iowa	1,423,567
Minnesota	109,016
	4,896,309

Or more than one-third.

The entire population of the United States in 1860 was 31,443,322.

Illinois contained..1,711,951
Iowa.. 711,951
Missouri...1,182,012
Minnesota.. 172,123
Wisconsin... 775,881

4,553,918

Or about one-seventh.

Thus it will be seen that these five States, possessing only one-seventh of all the population, and one-sixth of all the improved land, nevertheless, in 1864, produced more than one-fourth in value of the entire crop—more than one-fourth in value of all the live stock—more than one-third in number of all the cattle and hogs, and nearly one-half of all the wheat and corn grown in the United States. Here we find four and one-half millions of agriculturists, along the Upper Mississippi, producing, in a single year, from one-third to one-half of all the productions of the leading staples, of an estimated value of six hundred and seventy-seven millions fifty-six thousand two hundred and four dollars.

An examination of the statistics fully establishes the additional fact that these five States, during the years 1861, '62, and '63, shipped East 150 per cent. more corn and meal, and 25 per cent. more pork products, than were exported from the entire country during the same period. These States not only supply the export wheat of the entire country, but also the export corn and pork products. The contributions, therefore, made by Illinois, Wisconsin, Misssouri, and Minnesota, to the exports of the United States in these three leading agricultural staples alone, are as follows:

	1860-1.	1861-2.	1862-3.
Wheat............................	$48,938,780	$44,187,148	$55,647,979
Corn and meal................	6,387,160	9,609,879	9,623,357
Pork products................	4,687,784	10,217,281	16,424,338
Total............................	$60,013,724	$64,014,308	$81,695,674

The entire exports of domestic products of the United States amounted to—

1860-1.	1861-2.	1862-3.
$217,006,053...............	$190,699,387...............	$200,666,110

The average exports of the country for the three years were $222,874,183 33, and the average exports which these five States contributed in wheat, corn, and pork alone, was $68,575,568 66, very nearly one-third.

In 1861, '62, and '63, the average yearly tonnage of all American vessels engaged in trans-oceanic commerce, and entering the ports of the United States, was 2,564,257 tons, and the average tonnage of all the vessels of all countries engaged in oceanic commerce, and entering the ports of the United States, was 5,841,867 tons. Now, the three staples contributed by these five Upper Mississippi States to our exports were equivalent to 1,315,000 tons annually. They, therefore, not only contributed one-third in value to our entire exports, but gave employment upon the ocean to more than one-half of all our American tonnage, which was equivalent to one-fourth of all the tonnage of all nations, our own included, entering the United States, and engaged in trans-oceanic commerce. History cannot furnish a parallel.

The Agricultural Bureau, basing its calculation on past results, makes the following approximate estimate of the cereal product of the Northwest for the next four decades:

Years.	Bushels.
1870	762,200,000
1880	1,219,520,000
1890	1,951,232,000
1900	3,121,970,000

We consume in this country an average of about five bushels of wheat to the inhabitant, but, if necessary, can get along with something less, as we have many substitutes, such as corn, rye, and buckwheat. It is estimated that our population will be, in—

1870	42,000,000
1880	56,000,000
1890	77,000,000
1900, more than	100,000,000

Accordingly, we can use for home consumption alone, of wheat, in—

1870	210,000,000 bushels.
1880	280,000,000 "
1890	385,000,000 "
1900	500,000,000 "

From 1790 to 1817, breadstuffs were the chief exports of some of the New England and nearly all of the Atlantic States. Now New England produces but eleven quarts of wheat to each inhabitant, and consumes annually of agricultural productions $50,000,000 more than she produces. Pennsylvania, the first, and New York, the third, among the States in the production of wheat in 1860, are now calling upon the West, the former for ten per cent. and the latter for sixty per cent. of its bread; while Ohio, so long the promise land of the emigrant, is now growing but very little more wheat than will meet the wants of a population equal to her own. Nearly every State in South America, and nearly every nation in Europe, imports agricultural products, and in 1863 the United States sent its breadstuffs to sixty different foreign markets.

Russia, the chief grain exporting country of the Old World, from 1857 to 1862, inclusive, only exported annually:

Wheat...19,897,292 bushels.
Corn... 2,211,932   "

Thirty years ago steamboats engaged in the river trade aggregated but a few score. Now there are over a thousand. In 1865 the imports of St. Louis, Cincinnati, Louisville, and two or three minor Mississippi towns, were of the value of $730,000,000. As the export trade of these places was about equal to their imports, we have for the entire commerce of these points nearly $1,500,000,000. But this does not include the commerce of New Orleans, Memphis, Dubuque, and other important towns. Include the trade of these points, and the aggregate value of the trade of the Mississippi and its tributaries (the Ohio and Missouri) in 1865 was more than two thousand millions of dollars — a sum equivalent to three times the whole foreign commerce of the United States.

However important the above figures may appear, they must be taken as only a fraction of what will be the yield of the Valley States when they reach a high state of cultivation.

Not only are we great in coal, iron, wheat, and corn, but transcendent in the production of the precious metals, as the following tables will show:

COINAGE COMMERCE.—*Product of the whole American continent from its discovery, in 1492, to the commencement of 1868, a period of three hundred and seventy-six years.*

	Produce of the mines from 1492 to 1804.		From 1804 to 1848.		From 1848 to 1868.		Total value of each metal from 1492 to 1868.		Total value of both metals from 1492 to 1868.
	Gold.	Silver.	Gold.	Silver.	Gold.	Silver.	Gold.	Silver.	
United States			$25,000,000		$1,015,000,000	$100,000,000	$1,040,000,000	$100,000,000	$1,140,000,000
Mexico	$79,005,000	$1,948,952,000	62,000,000	922,560,000	50,000,000	120,000,000	191,060,000	3,041,452,000	3,232,452,000
South America	1,259,000,000	2,214,572,000	351,274,430	401,051,670	168,080,000	281,560,000	1,778,581,430	2,897,210,520	4,676,672,820
British America					37,000,000	1,000,000	37,000,000	1,060,000	38,060,000
Central America			8,860,000	4,400,000	5,000,000	3,000,000	13,860,000	7,460,000	21,320,000
Totals	1,348,500,000	4,163,520,000	447,674,430	1,077,981,670	1,255,080,000	505,560,000	3,060,654,430	6,087,671,420	9,107,725,820

Amount of both metals prior to 1801 .... $5,512,020,000
Amount of both metals since 1801 3,595,626,000

Difference $1,916,334,000

Average annual gold product from 1801 to 1843 .... $10,109,782
Average annual gold product from 1843 to 1868 62,254,000
Average annual silver product from 1804 to 1848 24,500,000
Average annual silver product from 1848 to 1868 40,278,000

Products of Gold and Silver from 1492 to the commencement of 1868, in America, Europe, Asiatic Russia, Australia, New Zealand, and portions of Northern Africa.

	Produce of the mines from 1492 to 1804.		Produce of the mines from 1804 to 1848.		Produce of the mines from 1848 to 1868.		Amount of each metal from 1492 to 1868.		Total amount of both metals from 1492 to 1868.
	Gold.	Silver.	Gold.	Silver.	Gold.	Silver.	Gold.	Silver.	
America..............	$1,348,500,000	$4,163,530,000	$117,954,430	$1,077,981,674	$1,263,680,000	$805,560,000	$2,660,654,430	$6,047,071,452	$9,107,725,882
Europe and Asiatic Russia............	363,800,000	291,600,000	457,400,030	145,300,050	340,000,000	100,000,000	1,165,200,000	537,400,000	1,702,600,000
Australia and New Zealand..........					800,000,000	5,500,000	800,000,000	5,500,000	805,500,000
	1,737,300,000	4,455,130,000	884,474,430	1,223,281,674	2,493,680,000	971,060,000	5,116,854,430	6,649,971,430	11,766,825,889
Existing in Europe in 1492........	60,000,000	140,000,000					60,000,000	140,000,000	200,000,000

The precious metals existing in Europe at the date of the discovery of America have been computed as amounting to $60,000,000 gold, and $140,000,000 silver, which, if added to the totals as stated in the above table, will produce $5,176,854,430 gold, $6,789,971,430 silver, and $11,966,825,889 in both metals, as the amount of gold and silver in use among civilized nations since 1492. If to these aggregates we add the somewhat conjectural product of Africa and Central and Southern Asia, for the last twenty years, we obtain, as a grand total, $5,656,854,430 gold, $7,029,971,430 silver, and $12,966,825,889 as the amount of both metals, appropriated to the uses of mankind during the last 376 years.

It will be seen by the above figures that from 1848 to 1868 our gold mines have yielded more than those of any other country on this hemisphere. And it will be seen by the following table that the yield of North and South America together, since 1492, surpasses that of any other nation on the earth.

TABLE *showing the growth of coinage of the United States from 1793 to 1867.*

YEARS.	Gold.	Silver.	Copper.	TOTAL.
1793 to 1800, 8 yrs	$1,014,290 00	$1,440,454 75	$79,390 82	$2,534,135 57
1801 to 1810, 10 yrs	3,250,742 50	3,569,165 25	151,246 39	6,971,154 14
1811 to 1820, 10 yrs	3,166,510 00	5,970,810 95	191,158 57	9,328,479 52
1821 to 1830, 10 yrs	1,903,092 50	16,781,046 95	151,412 20	18,835,551 65
1831 to 1840, 10 yrs	18,791,862 00	27,199,779 00	342,322 21	46,333,963 21
1841 to 1850, 10 yrs	89,443,328 00	22,226,755 00	380,670 83	112,050,753 83
1851 to 1860, 9½ yrs	470,838,180 98	48,087,763 13	1,249,612 53	520,175,556 64
1861 to 1867, 7 yrs	296,967,464 63	12,638,732 11	4,869,350 00	314,475,546 74
Total, 74 yrs	$885,375,470 61	$137,914,587 14	$7,415,163 55	$1,030,705,141 30

I have already intimated that, instead of the people of the Valley States looking to India, China, and Japan, for commerce, as the popular but superficial judgment seems now to incline, their interest lies in the tropics of our own hemisphere; that instead of indulging in wild and chimerical speculations, across distant oceans to distant lands, for things relatively useless in life, they must look to the islands of the Gulf, Mexico, Central and South America, for the wealth and products of those countries.

An important element of that wealth may be inferred from the following table:

CHANGE OF NATIONAL EMPIRE. 153

Produce of the South American mines from the discovery of the continent to the end of the year 1867.

	From the discovery of the continent to the end of the year 1803.		From 1804 to the end of 1847.		From 1848 to the end of 1867.		Total value of each metal since 1803.		Total value of both metals since 1803.
	Gold.	Silver.	Gold.	Silver.	Gold.	Silver.	Gold.	Silver.	
Peru...........	($416,000,000	$2,214,578,000	($26,400,000	$266,800,000	$12,000,000	$94,000,000	$438,400,000	$2,300,300,000	$3,338,300,000
Bolivia........			19,614,430	117,789,215	8,920,000	36,000,000	28,534,430	173,789,660	202,323,430
Paraguay and Uruguay..			3,300,000	18,700,000	1,500,000	8,500,000	4,800,000	27,200,000	32,000,000
Argentine Confederation...			3,300,000	18,700,000	4,540,000	25,500,000	7,860,000	44,200,000	52,060,000
Chili..........			56,760,000	38,620,450	15,160,000	90,000,000	71,920,000	128,630,450	195,550,450
New Granada...			140,800,000	170,000	65,000,000	5,200,000	205,800,000	5,370,000	211,170,000
Ecuador and Venezuela...			13,100,000	100,000	8,000,000	2,000,000	21,100,000	2,100,000	23,200,000
Brazil.........	855,500,000		89,000,000	732,000	40,000,000	920,000	128,000,000	1,152,000	129,152,000
Total........	1,268,500,000	2,214,578,000	351,274,430	401,081,674	158,080,000	281,560,000	909,354,430	682,641,430	1,191,995,880

Total amount of gold from 1492 to 1868, $1,778,854,430; silver, $2,897,219,430; both metals, $4,676,073,860.
Gold product from 1804 to 1868, $509,354,430; silver product, $682,641,674; both metals, $1,192,000,000.

The present annual product of the several divisions of South America may be computed as follows:

	Silver.	Gold.
Peru............	$5,000,000	$600,000
Bolivia.........	2,800,000	450,000
New Granada...	200,000	3,400,000
Chili............	4,500,000	750,000
Brazil..........	18,000	2,000,000
Paraguay and Uruguay...	425,000	75,000
Argentine Confederation...	1,275,000	225,000
Ecuador and Venezuela...	100,000	400,000
Total........	$14,578,000	$7,900,000

By the preceding tables it will be seen that a people grasping for wealth can find more of it in the Southern countries than in the Orient, and certainly they will go there for it.

ISTHMUS CANAL.

Besides whatever importance there may be attached to the trade of the Orient, our practical age, united with the necessities of our commerce, will demand the construction of a ship canal across the Isthmus of Darien. Not only our own commerce, but also that of England and France, demand its construction. The most important interest demanding its construction, at the present time, is represented by the following tables:

Table of the saving in distance from New York to the following places, by the Isthmus of Panama, over the Cape routes.

From New York to—	Distance via Cape of Good Hope.	Distance via Cape Horn.	Distance via Isthmus of Panama.	Saving in distance over the route by Cape Horn	Saving in distance over the route by Cape of Good Hope.
	Miles.	Miles.	Miles.	Miles.	Miles.
Calcutta	17,500	23,000	13,400	9,600	4,100
Canton	19,500	21,500	10,600	10,900	8,900
Shanghai	20,000	22,000	10,400	11,600	9,600
Valparaiso	....	12,900	4,800	8,100	....
Callao	....	13,500	3,500	10,000	....
Guayaquil	....	14,300	2,800	11,500	....
Panama	....	16,000	2,000	14,000	....
San Blas	....	17,800	3,800	14,000	....
Mazatlan	....	18,000	4,000	14,000	....
San Diego	....	18,500	4,500	14,000	....
San Francisco	....	19,000	5,000	14,000	....
Wellington, N. Z	13,740	11,100	8,480	2,620	5,260
Melbourne, Australia	13,230	12,720	9,890	2,830	3,340

CHANGE OF NATIONAL EMPIRE. 155

Table showing the trade of the United States that would pass through the Isthmus Canal if now finished.

Countries traded with.	Exports and Imports.	Tonnage.
Alaska	$ 126,537	$ 5,735
Dutch East Indies	904,550	16,589
British Australia and New Zealand	4,728,083	52,105
British East Indies	11,744,151	177,121
French East Indies	98,432	3,665
Half of Mexico	9,601,063	34,673
Half of New Granada	5,375,354	131,708
Central America	425,081	36,599
Chili	6,645,634	63,749
Peru	716,679	193,131
Ecuador	48,979	1,979
Sandwich Islands	1,157,849	33,876
China	12,752,062	123,578
Other ports in Asia and Pacific	80,143	4,549
Whale Fisheries	10,796,090	116,730
California to East United States	35,000,000	861,698
Value of cargo	$100,294,687	$ 1,857,485
Value of ships at $50 per ton	92,874,250	
Total value of ships and cargo	$193,168,937	$92,874,250

Table showing the trade of England that would pass through the Isthmus Canal if now finished.

Countries traded with.	Exports and Imports.	Tonnage.
Half of Mexico	$ 2,775,137	$ 11,833
Half of Central America	1,244,817	5,615
Half of New Granada	2,437,605	10,188
Chili	15,486,110	118,311
Peru	20,473,520	244,319
Ecuador	360,015	1,820
China, } Outward only, forty days saved by Canal	7,077,390	68,520
Java,	3,821,410	16,003
Singapore,	4,364,070	16,500
Australia and New Zealand	78,246,695	522,426
Sandwich Islands	520,560	1,950
California	2,378,105	11,800
Value of trade	$139,184,834	$ 1,029,295
Value of ships at $50 per ton	51,464,750	
Total value of trade and ships	$190,649,584	$51,464,750

Table showing the trade of France that would pass through the Isthmus Canal if now finished.

Countries traded with.	Exports and Imports.	Tonnage.
Chili	$10,000,000	25,688
Peru	13,160,000	
Half of Mexico	2,790,000	10,004
Half ot New Granada	1,090,000	2,389
Ecuador	440,000	1,651
Bolivia	100,000	1,000
California	2,073,859	8,997
China, } Outward only	2,180,000	2,028
Dutch East Indies,	4,440,000	20,400
Sandwich Islands	2,000,000	4,119
Phillipine Islands	1,000,000	1,463
Australia	19,800,000	50,000
Value of cargoes	$59,073,859	162,735
Value of ships at $50 per ton	8,136,750	
Total value	$67,210,609	8,136,750

Table showing the total tonnage that would yearly pass through the Isthmus Canal if now finished.

	Tons.
United States	1,857,485
England	1,029,295
France	162,735
Other countries	44,555
Total	3,094,070

Table showing the general result of the foregoing tables.

Tonnage and trade of the United States			$193,168,937
" " " England			190,649,584
" " " France			67,210,609
" " " Other countries			16,802,000

These tables show items of vast importance to the trade of the world and its approaching change. That a ship canal will be constructed across the Isthmus, there is no manner of doubt. In further demand of its construction, the United States would save yearly by it, in her shipping, $35,995,930. England would save by it yearly $9,950,348. France would save by it yearly $2,183,930. The trade of the world would save by its construction $49,530,208.

With this vast amount of trade awaiting its construction, we can safely say that the time for its completion is not remote, and it will give a new growth and vitality to our country and to the Valley States. Then will the Mississippi river and the Lake and Gulf railways become the greatest commercial channels on the continent, thus increasing our internal trade and augmenting the commercial supremacy of St. Louis. The close proximity of our Gulf ports to such a canal would necessarily control that portion of our government trade for the Valley States, and, by the necessities of our trade and the wants of the people, give the control of that trade to St. Louis.

Several efforts have been made within the last sixteen years to provide for the construction of this great canal, but as yet without success. Very recently a meeting was held in New York city for the purpose of organizing a company and securing the required financial aid for its construction. The following statement of the meeting is taken from the December number of *Appleton's Railway Guide*, and will be found interesting to the reader:

IMPORTANT COMMERCIAL ENTERPRISE—A CANAL TO BE MADE ACROSS THE ISTHMUS OF PANAMA.

At a meeting of the corporation of the Isthmus Canal Company, held at the residence of Peter Cooper, Esq., the company was organized by electing Mr. Cooper as president, and Fred. A. Conklin as secretary.

The Secretary of State of the United States, the Hon. William H. Seward, and the Attorney-General of the United States, the Hon. William M. Evarts, having come from Washington to confer with the leading capitalists and merchants of this city upon the subject, were present, and laid many important facts before the meeting. Estimates from the highest sources state the cost of the work at $100,000,000. The following gentlemen were appointed commissioners to obtain subscriptions to the stock of the company: W. F. Coleman, Marshall O. Roberts, Cornelius K. Garrison, William B. Duncan, and Richard Schell.

Among those present at the meeting were many gentlemen prominent as capitalists, merchants, and as members of the learned professions. Charts of surveys of the proposed route, by Frederick W. Kelley and other eminent engineers, were exhibited, which demonstrated the feasibility of the undertaking, and entire confidence was expressed in its ultimate success as a work of engineering and as a commercial enterprise.

During the discussion, the Hon. William H. Seward spoke substantially as follows:

SPEECH OF THE SECRETARY OF STATE.

"GENTLEMEN: Ever since the canal of the Pharaohs across the Isthmus of Suez fell into disuse, and was lost under changes of society and nature, commerce has desired the restoration of that original and most feasible channel of trade and intercouse between the Atlantic and the Pacific nations. The discovery of the Cape of Good Hope supplied a costly and hazardous substitute, which was eagerly accepted. The exploration of the newly-discovered American continent, at the beginning of the sixteenth century, disclosed at once necessities for a better channel to be constructed across that continent, and made a full revelation that that better channel could be constructed across that continent, and nowhere else. During the past three hundred years statesmanship and humanitarianism have combined with ever-increasing diligence and effort to find the means of effecting an enterprise which is, perhaps, the only one that ever has commanded universal assent and commended itself to the desire of all mankind. Every advance of modern civilization in Europe, the establishment of every new nation in America, every opening of any secluded Asiatic State and nation that has occurred, has increased the zeal and the energy of the friends of progress in favor of a canal across the Isthmus of Darien. We habitually feel and say that we are living in an important and interesting period. We do, indeed, have occasion and opportunity to labor effectually in various ways in the cause of civilization and humanity; but, if I do not mistake, the chief of all the advantages of statesmen of the present day in all the countries is, that they can take part in the construction of a canal across the Isthmus of Darien.

"Gentlemen, to accept our respective parts in this great enterprise is the work of this night. We are Americans. We are charged with responsibilities of establishing on the American continent a higher condition of civilization and freedom than has ever before been attained in any part of the world. We all acknowledge and feel this responsibility. The destiny which we wish to realize as Americans is set plainly before us, and distinctly within our reach; but that destiny can only be attained by the execution of the Darien ship canal. The reason is obvious. While the electric telegraph can and must be used for the interchange of ideas between nations, and while improved highways must and will be used for overland travel and intercourse, yet the mineral, forest, and agricultural bulky productions of the earth can only be exchanged by navigation, and this navigation must be made as cheap and as frequent and as expeditious as is possible. But as to navigation by sailing vessels, commerce can no longer afford to us the circuitous and perilous navigation around the Capes. It must and will have shorter channels of transport, and of these there can be but two—the one across the Isthmus of Suez, the other across the Isthmus of Darien. A canal across the Isthmus of Suez already approaches its completion. If that channel is to secure the patronage of universal commerce, it will be fully enlarged and completely adapted to the interests of modern commerce. In that case the commerce of even the Atlantic American coast, from the St. Lawrence to Cape Horn, will be turned eastward across the Atlantic, and through the Mediterranean and Red Seas, and the Indian Ocean, to India and China. It would be a reproach to American enterprise and statesmanship to suppose that we are thus to become tributaries to ancient and effete Egypt, when by piercing the Isthmus of Darien we can bring the trade of even the Mediterranean, and of the European Atlantic coast, through a channel of our own, so palpably indicated by nature that all the world has accepted it as feasible and necessary.

"We have undertaken to develop the resources of our own continent, and to regulate and restore the Asiatic nations to free self-government, prosperity, and happiness. The Darien ship canal is the only enterprise connected with the great work of civilization which remains to be undertaken. It was a mistake to suppose that we have been hitherto either inactive or idle in regard to this important matter. We have built a railroad across the Isthmus of Panama, and within twelve months more we shall have stretched a railroad across the continent from New York to San Francisco. We have abundant assurance that these achievements are profitable and useful. Both of them, however, are profitable and useful only as types and shadows of the Darien ship canal, which we all feel and know must be transcendently profitable and transcendently useful.

"The executive Government of the United States, gentlemen, has adopted the enterprise in which you are engaged. It has provided for a full, satisfactory, and final survey, preparatory to the construction of the Darien ship canal. It is engaged in negotiating with the Republic of Colombia for its consent to your achievement of the enterprise. The President will go forward with renewed zeal and vigor on receiving the assurances which you have given me that the city of New York has named the men who will undertake that achievement, and stand ready to furnish the hundred million dollars which it may be expected to cost."

EXPENSES.

Lest some should still be ready to object on account of some trivial cause, and especially on account of the public expense necessary to make the change and erect new buildings, I submit the following statement from the Hon. Hugh McCulloch, Secretary of the United States Treasury, giving the cost of the public grounds and buildings at Washington:

TREASURY DEPARTMENT, September 28, 1868.

SIR: In reply to your inquiries, I have to say that the total amount expended in the District of Columbia from the time the seat of government was located there to June 30th, 1868, for public works of every description, including buildings and works of art, is $37,390,853.08.

The real estate, exclusive of buildings, was assessed at $13,412,293.26, in 1858. Since that time there has been no assessment of which the Department is advised.

Very respectfully,
HUGH McCULLOCH,
Secretary of the Treasury.

L. U. REAVIS, ESQ.,
St. Louis, Mo.

By this statement it will be seen that the expense of erecting the public buildings at Washington City is far below the amount supposed by those who have ventured an opinion upon the subject.

On the following page will be found a statement from the Hon. O. H. Browning, Secretary of the Interior, showing the extent of the public grounds in the District of Columbia. The statement, like that of the public expenditures, is much below that which the uninformed individual would have estimated.

CHANGE OF NATIONAL EMPIRE.   161

DEPARTMENT OF THE INTERIOR,
WASHINGTON, D. C., December 19, 1868.

Sir: Referring to your request of the 1st inst., I inclose herewith for your information a statement showing the extent of the public grounds in the District of Columbia.

I am, sir, very respectfully,
Your obedient servant,
O. H. BROWNING,
Secretary.

L. U. REAVIS, ESQ.,
St. Louis, Mo.

STATEMENT OF RESERVATIONS,

PROPERTY OF THE UNITED STATES, IN THE CITY OF WASHINGTON, D.C.

SHOWING AREA OF EACH.

No.	DESIGNATION.	Square feet.	Acres.
1	President's grounds (including White Lot)	3,516,721	80.7344
	Lafayette Square	304,964	6.9948
2	Capitol Square and Mall	2,121,141	48.6946
	Mall bet. 3d and 4th sts., west, and Missouri av. and Canal	88,175	2.0242
	Mall bet. 3d and 4th sts., west, and Maine av. and Canal	88,967	2.0424
	Mall bet. 4½ street, west, and Missouri avenue	333,212	7.5806
	Mall south of Canal, bet. 4½ and 7th sts. (Armory Square)	751,328	17.2483
	Mall bet. 7th and 12th sts., west, (Smithsonian grounds)	2,289,192	52.5525
3	Mall bet. 12th and 14th sts., west, (Agricultural Departm't)	1,378,172	31.6384
4	Mall west of 14th street to Potomac river	1,956,789	44.9217
	Observatory Square	889,674	19.2625
		Square feet.	Acres.
5	Arsenal grounds, original grounds... 1,249,000	28.6637	
6	Arsenal grounds, recent purchase... 1,805,017	41.4379	3,054,917	70.1316
7	Western Market		
8	Center	161,209	3.7006
9	Patent Office (Interior Department) Square	181,221	4.1603
13	Judiciary Square	836,374	19.2005
14	Hospital Square, on Eastern Branch	3,361,199	77.1627
15&16	Navy Yard	1,871,333	42.9000
17	Eastern Market	112,878	2.5913
	Town House Square	1,036,191	23.7876
	Maine Barracks, Square No. 927	156,179	3.5853
	Franklin Square, Square No. 249	174,418	4.0040
	General Post Office, Square No. 430	64,300	1.4761
	Total	25,189,002	578.0026

A change of the seat of government does not imply a loss of the public buildings at Washington, by any means, for all the valuable material can be easily moved and put in the new buildings; and by this means new and better buildings can be erected at a less cost than were the present ones. Good engineers freely express the entire feasibility and safety of taking down and removing either or all the present government buildings, and they can be taken down, moved, and newly erected, in five years. Boats can bring the materials all the way round by water, and land them at the new seat of government at far less expense than was first required to collect them together at Washington.

But the cost of the change and the erection of the buildings is a matter of but small consideration. An expense of $10,000,000 would be of little concern to this great nation, and especially when it would probably be twenty years in spending it. It is true that for the Capital of the New Republic would be required buildings of more magnificent structure than those of the Old Government — more magnificent than were ever yet wrought by human hands. In anticipation of loftier and purer American statesmen than now are, the Republic will require more magnificent legislative halls. In anticipation of the future grandeur and goodness of the Republic, department buildings far superior and more commodious than the present will be required. In anticipation of a wiser and better people all over the land, the New Republic will be required to give national aid to the distribution of knowledge among its citizens and mankind, and thus will be demanded departments for these beneficent purposes. Yet the expense for all is insignificant, when considered in the light of the future growth of the Republic.

Again, the national expense will be reduced, by the removal of the seat of government to the Valley States, by cutting short the mileage of new members of Congress that will yet claim seats as representatives and senators of the new States yet to be born into the family of the Republic. This item alone, small as it may seem, will in time show largely on the side of economy.

Again, there exists an intolerable objection to the seat of

government remaining at Washington, on account of the inconvenience to reach it, and also on account of the poverty and monopoly of its markets. Those who know anything of the markets at Washington City know that they are scantily supplied, and that, too, with products that bear no comparison with Western products; and besides their inferiority and scarcity, the people are compelled to buy or do without. These would seem to be insignificant items; but when we consider the immense use of such products at a national capital, they at once become items of great concern.

The seat of government, located at St. Louis, will be placed at the center of the best means of public communication from all parts of the country afforded on the continent; besides, situated in the midst of the best products of the Mississippi Valley, where there can be no scarcity and no monopoly.

It has been foolishly argued by some that the seat of government, at any point, is a means to generate demoralization and corruption in the people. This objection is so silly that it deserves to be noticed in order to render it contemptible. It is one of those objections often made by individuals who can always see more faults in their neighbors than they can in themselves. It is made by those who look upon the dark side of the picture of human life with doubt and distrust, and, by thus expressing themselves, are enemies to the highest interests of human society. Away from that dark picture; away from the faith or influence of him or her that does not have implicit confidence in the success of the Republic! Never before in the history of mankind has individual or national life reached so high a plane in intellectual and moral progress as at the present hour. To contend that the seat of government of the Republic is a means to breed corruption and guilt, is to contend that one's self is a villain, and that his neighbors are hypocrites and demagogues, that society is a farce, and the law a blank. It is not so. The capitals of England, France, Turkey, China, Russia, Mexico, and nearly all the great nations of the earth, are located at the great cities.

Let us have faith in the people, and let it be said in all truth that if the people send honest and upright men to the national legislature, society will be as pure and statesmanship as elevated at the seat of government as the most upright and enlightened can desire.'

Let us look on the bright side, and resolve that none shall represent the New Republic but the pure and the wise, the faithful and the upright, and all will be well.

One of the most important duties for the American people to perform is that which looks directly to the elevation of the national life, and this work must be begun at home. Let no man be blind to this fact. If the stream is impure, the fountain must be, also. If the people want temperance, virtue, morality, honesty, and moral and intellectual grandeur, in city councils, State legislatures, and in the national Congress, they must first acquire the supremacy of those excellencies at home; and they who do not contend earnestly for these virtues and attainments at home are hypocritical grumblers against their neighbors and rulers. Then, let the lesson first be unerringly taught at home, and its meaning will be indelibly impressed upon the national life.

Already the sentiment for a better state of society and government is germinating in the hearts of the people, and corrupt politicians will soon give place, all over the land, to worthy and capable statesmen. Let us all labor to hasten the change, in the hope that, when some future Plutarch weighs the coming men of the Republic, they will be the grandest growth of the human race.

SPECIAL AND LOCAL CONSIDERATIONS.

In addition to the general arguments which have been given in the preceding pages in favor of the removal of the seat of government to the Mississippi Valley, and the indication of St. Louis as the most suitable place for it, the following map of lands, together with local and special facts, are presented as supplemental considerations.

Last winter the Hon. C. A. Newcomb, member of Congress from Missouri, offered a bill in Congress providing for the removal of the seat of government from Washington City to St. Louis. In co-operation with Mr. Newcomb's bill, the Hon. G. A. Finkelnburg, member of the Legislature of Missouri, offered the following bill authorizing the State of Missouri to cede a certain portion of her territory to the exclusive use and control of the General Government, in consideration of the National Capital being moved to the portion of territory ceded.

AN ACT to cede a portion of the territory of St. Louis county, in the State of Missouri, to the United States of America, for a seat of government of the United States.

SECTION 1. *Be it enacted by the General Assembly of the State of Missouri, as follows:* That so much of the territory of the State of Missouri as is included within limits and boundaries following, to-wit: Beginning at a point in the Mississippi river one mile south of Chouteau avenue, in the city of St. Louis; thence west ten miles; thence south to the center of the Meramec river; thence down the Meramec river to its junction with the Mississippi river; thence up the Mississippi river to the place of beginning — be and the same is hereby ceded and transferred to the United States of America; and all the right, title, authority, and jurisdiction, now owned, possessed, exercised, and enjoyed, by the State of Missouri, in or to or over said territory, is hereby vested in the United States of America, upon the sole and express condition that the seat of government of the United States of America shall be removed to said territory on or before the first day of January, 1880.

SEC. 2. *Be it further enacted,* That said territory shall not vest in the United States of America until Congress shall pass an act moving the seat of government of the United States of America to said territory, and authorizing the laying out of a capitol and public grounds; and any removal of the seat of government from said territory any time thereafter shall immediately revert all the title, jurisdiction, and authority in, to, and over said territory in the State of Missouri: *And provided, further,* that no change of government, or jurisdiction, which may take place under this act, shall affect the rights of property of individuals or bodies corporate within the territory aforesaid.

SEC. 3. As soon as Congress shall pass an act removing the seat of government to said territory, the Governor shall formally transfer the same to the United States of America.

SEC. 4. The Governor shall forward copies of this act to the presiding officers of each House of Congress, to be laid before said Houses for consideration.

The district described in the above bill has an area of about 90 square miles.

A vote was taken in the House of Representatives on Judge Newcomb's bill, and, under the circumstances, was more favorable for the removal than the friends could have expected. Owing to the lateness of the time at which Mr. Finkelnburg's bill was introduced in the Legislature, a vote was not reached. But there can be no question about the State of Missouri ceding to the General Government such a district of territory as may be required for the purposes of a National Capital.

SHAW'S SUBDIVISION.

The map of grounds submitted to illustrate this part of the subject of the pamphlet shows the district described in Mr. Finkelnburg's bill, with an addition of a strip one mile in width on the north side, which is added to include Mr. Shaw's subdivision and his splendid garden.

Mr. Finkelnburg made the selection of the district described in his bill, in accordance with the public sentiment of the people of St. Louis, who, without hesitation, look thither to several beautiful sites, one of which seems fated to be the seat of empire for the New Republic.

By reference to the map of this district, it will be seen there are four shaded tracts of land, three lying upon the Mississippi river, and one back from it. The two southern tracts of land, as will be seen on the map, are the property of the Hon. Henry T. Blow, and known as Clifton Heights. The tract farthest up the river is Jefferson Barracks, and is owned by the Government. The shaded tract on the north side of the map, and out from the city, is the property of Mr. Henry Shaw, whose reputation has gone all over the country, on account of his fine botanical garden. Mr. Shaw's property is situated at a distance of four and five miles from the river. His garden property and that immediately around it, although extremely beautiful and valuable, is not so elevated as many of the adjoining locations. However, beyond his garden, a little more than one mile, is a beautiful, broad prairie ridge, with an elevation of 200 feet above the city directrix. It is the most beautiful site back from the river that is in the vicinity of the city of St. Louis, and on account of its situation, its elevation, and its surroundings, it is regarded as one of the favorite and suitable localities for the Government to erect new public buildings. Besides the general favorableness of this location, its close proximity to Mr. Shaw's fine garden would be an item of great concern to the new seat of government, for his is much the finest garden in the United States, and infinitely surpasses those of the Government at Washington. Land adjoining Mr. Shaw's sub-division is from $500 to $2,000 per acre.

CLIFTON HEIGHTS.

Passing from Mr. Shaw's sub-division to Clifton Heights, the property of Mr. Blow, we find an entirely different situation.

Clifton Heights is situated ten miles below the city. It consists of 1,300 acres, and, as will be seen by the shadings upon the map, lies immediately upon the great Father of Waters, and has a river front of five miles. The topography and general character and location of Clifton Heights cannot be equaled by any other property upon the river in the vicinity of St. Louis, and is not surpassed anywhere from its source to its mouth. It is high and commanding, with views equal to any of the Alleghanies; and when the art of man, with equal skill, there unites with nature, no people can point to a place more remarkable for its beauty, healthfulness, and commanding position.

In fact, it would seem that nature, far back in the past, had especially provided this beautiful site on the great Mississippi river for the seat of empire for the New Republic.

This property has an elevation of 252 feet above the river, which is much higher than any other ground around the city, and from its favorable situation it commands a view of 50 miles each way up and down the stream; and with the Government buildings erected upon this elevated plateau, they will occupy a position which, united with their great size, will afford a view to the traveler upon the river or railroad which will far surpass any views on the Hudson or Potomac — views which will never grow dull to the vision.

Land about Clifton Heights ranges from $80 to $300 per acre, and every advantage for the supply of good water and good building stone is afforded from the bluffs of the Mississippi and the Meramec rivers.

JEFFERSON BARRACKS.

Situated on the river, just above Clifton Heights, is Jefferson Barracks, the property of the Government, consisting of 1,700 acres, which would afford ample room for the Government buildings, but for its unfavorable character would not be so well suited for such an important purpose. The ground is not so elevated as that of Clifton Heights, or as that of Mr. Shaw's division. Therefore it is not probable that the Government would fix upon that tract in the event of a change, but would, no doubt, retain it for a few years, until the sale of it would be an item of pecuniary importance.

In this connection it may be proper to state that when Congress orders a removal of the seat of government, it will be necessary, as under the Old Government, to provide for the temporary use of buildings for Congress and the departments at the place selected for the new Capital, in order to admit of the removal of the present ones at Washington. In that event, St. Louis, or whatever other place may be selected, will no doubt be asked to furnish suitable buildings for temporary use.

WHAT TIME.

> The general mind is faithless of what goes much beyond its own experience. It refuses to receive, or it receives with distrust, conclusions, however strongly sustained by facts and fair deductions, which go much beyond its ordinary range of thought. It is especially skeptical and intolerant toward the avowal of opinions, however well founded, which are sanguine of great future changes. It does not comprehend them, and therefore refuses to believe; but it sometimes goes further, and, without examination, scornfully rejects. To seek for the truth is the proper object of those who, from the past and present, undertake to say what will be in the future, and, when the truth is found, to express it with as little refferrence to what will be thought of it as if putting forth the solution of a mathematical problem.—*J. W. Scott.*

The reader of this little pamphlet will no doubt be desirous to know what time the seat of government will be moved from its present place to the Mississippi Valley, or, at least, will be anxious to know what time one so sanguine as the writer has fixed for the change. I unhesitatingly answer that the change will be made within five years from January 1, 1869. Before two years from January 1, 1869, Congress will authorize, by its own act, the removal of the seat of government from its present place, and soon will follow the President, national archives, and the legislature of the Republic.

I know there are those who will regard this statement more visionary than any preceding one I have made; but, to such as choose to look with discredit upon it, I can only hope that experience will teach them that which they are now unable to comprehend. He who does not comprehend the workings of the under life-current of the Republic at the present time, is shut out from a comprehension of the future, and thus he becomes a conservative, a fogy—drift-wood in the rolling tide of progress.

Ours is a moving time. Changes come much sooner than most men expect them. The Hon. Horace Greeley but a few years ago did not expect slavery to be abolished in this century. Pof. Morse did not expect the ocean to be spanned by a telegraph for two or more generations hence. Dr. Lardner, the most learned philosopher of England in his day, declared in a lecture, in Liverpool, that he would eat the first steam engine that propelled a vessel across the ocean. Six months afterward a vessel did cross the ocean by the use of a steam engine, but was not eaten. There are, no doubt, some wise men with large stomachs, who will read this pamphlet or hear of it, that will propose to eat the first public building erected on the new Capital grounds in the Mississippi Valley in the next generation. There are, no doubt, men of would-be public spirit and enterprise who will readily volunteer to do this eating.

What is there to retain the Capital where it is? But two things—the local interests of the people of Washington City, and the consideration on the part of the Government of the public buildings erected at that place. I have already shown that the consideration of the public buildings at that place is an item of small consequence to the great and growing interests of the Republic. The local interests of the people of Washington City can have no weight in the matter whatever. It is purely a national question, and the representatives of the people must alone view it as such.

It is of no value whatever to New York to have the Capital at Washington. James Gordon Bennett, some years ago, declared that he would not give the patronage of the washer-women of New York for all the Government patronage. So, too, might the city of New York say, for she stands above Washington. None of her interests are subservient to Washington; therefore she will be unconcerned about the change.

Let me repeat again: the change will be made in five years, and before 1875 the President of the United States will deliver his message at the new seat of government in the Mississippi Valley.

A CHANGE

OF

NATIONAL EMPIRE;

OR

ARGUMENTS IN FAVOR OF THE REMOVAL

OF THE

NATIONAL CAPITAL FROM WASHINGTON CITY

TO THE

MISSISSIPPI VALLEY.

(Illustrated with Maps.)

BY L. U. REAVIS.

Fair St. Louis, the future Capital of the United States, and of the Civilization of the Western Continent.—JAMES PARTON.

There is the East, and there is India.—BENTON.

ST. LOUIS:
PUBLISHED AND FOR SALE BY J. F. TORREY, BOOK AND NEWS DEALER.
1869.

www.ingramcontent.com/pod-product-compliance
Lightning Source LLC
Chambersburg PA
CBHW020252170426
43202CB00008B/338